Storing
Home Grown
Fruit & Veg

Harvesting, preparing, freezing, drying, cooking,
preserving, bottling, salting, planning, varieties

Carol

foulsham
LONDON • NEW YORK • TORONTO • SYDNEY

foulsham

Capital Point, 33 Bath Road, Slough, Berkshire,
SL1 3UF, England

Foulsham books can be found in all good bookshops and direct
from www.foulsham.com

ISBN: 978-0-572-03630-0

Cover photographs © Garden Picture Library (top); James Murphy Photo
Library (bottom left, right and centre)

A CIP record for this book is available from the British Library

Printed in Great Britain by CPI Cox & Wyman, Reading, RG1 8EX

Contents

Introduction   5

PART ONE: Planning, harvesting and preserving

1 Planning for the larder   8
2 Gathering in the harvest   12
3 Storage   13
4 Drying   19
5 Freezing   27
6 Bottling   39
7 Jams, jellies, curds & butters   45
8 Pickles, chutneys & sauces   51
9 Juices & cordials   55
10 Cider & wine   59

PART TWO: A to Z of storing fruit & vegetables   67

Index   205
Useful websites   208

Introduction

Whether you have a large vegetable plot or a small one, or an orchard or just one apple tree, there are going to be seasons when a particular crop, or perhaps all your produce, is ready to eat at once. Of course it is very satisfying to be able to give away some of it to neighbours and friends, but after all the hard work you have put into sowing and nurturing your seeds to fruition, it would be better to be able to enjoy the fresh fruits of your labour throughout the year.

This book is packed with ideas on how to store your bountiful harvest, whether it be apples, beans or potatoes. However, to get the best out of each vegetable or fruit for as long as possible, it is important to choose the correct way of storage so that nothing goes to waste. Some produce, such as raspberries, freezes beautifully, whereas others, such as carrots, keep well packed in sand. Some may do better if they are cooked first and then frozen or made into preserves, juices or wine; or some may be suitable for drying or bottling, like our forebears used to do before the freezer became a household necessity. This book shows you all the alternatives to suit each vegetable and fruit.

There is little more satisfying than growing your own food and eating it. This way you can ensure that what reaches your table is completely fresh and full of health-giving vitamins and minerals. Your own fresh produce can also be free from chemicals, if you choose, will not have been genetically modified, or have clocked up any mileage in lorry transport; and because you can eat it straight from the plot, it will taste delicious.

Growing your own can be cost effective, too, as long as you don't have to hire anyone to look after your plot. Now that local councils are encouraging composting in their recycling plans – some offer compost bins at reduced prices – there is no excuse not to use your own home-made compost to create the ideal growing conditions for your produce that will result in a top quality harvest.

A complete and practical guide to storing your own fruit and veg, this comprehensive book shows how easy it can be to make the most of your harvest, from planning what varieties store best and how much to grow, to when is the best time to harvest and store, and the many different ways in which to do it. Part 1 describes step-by-step the different ways that you can store your produce to get the most from each crop – drying, freezing, juicing, bottling, preserving and making wine. The A to Z format in Part 2 looks at the most popular crops grown in vegetable plots and allotments today, with a few unusual ones, too and gives the best storage methods for each one, marked with a symbol for easy cross-reference to Part 1. Basic recipes are given for preserves, juice and wine making, with some extra mouth-watering ideas for cooking your produce for the freezer. With this book as your guide you will never need to waste your precious produce again – although your neighbours might not be so happy ... !

Part 1

Planning, harvesting & preserving

1 Planning for the larder

You will get the best results from your garden or allotment if you give a little time and thought to what you are going to grow and how you will plan your growing schedule.

What to grow

During the cold, unproductive months of winter, take time to look through the seed catalogues and plan what you are going to grow. Choose produce that is known to store well without losing any flavour, and when it comes to sowing and planting out, try to do it in succession – one batch every two to three weeks – so that not everything comes to fruition at once.

What not to grow

We have included information on how to store almost anything you can grown in a temperate garden so the only constraints on what you can grow are those you impose yourself. If you can eat your produce fresh, that's great. Then if you can barter your excess or, if not, store it in one way or another, then there's no reason not to grow it!

Aiming for freshness

The main aim is to store your harvest when it is at its freshest, but if everything is ready at once it is unlikely you will have enough time to get it all done as quickly as may be necessary. Also, the main pleasure in growing your own is to be able to eat it fresh from the plot, so it is not a good idea for all your crops to be ready for harvesting at the same time.

Successive varieties

With the right conditions, a glut of one vegetable or another is always a possibility, and of course with fruit trees, there are good years and bad years, but they are always ready during a short period. With soft fruit, different varieties span the growing season, so you can plant an early variety of, say, raspberry to fruit in July, another to fruit in August, and a third to be ready as late as October.

Storage space

This is an important aspect to consider when planning what to grow. Do you have the freezer space to store a glut of courgettes or tomatoes? The cupboard space for jars of pickles, conserves or bottles of wine? The shed space for the dry storage of apples, carrots or potatoes? Some vegetables such as leeks and root vegetables don't need to be stored as they can stay in the ground until you want them, but you will still need plenty of space in the vegetable plot, so that they don't get in the way of the next season's crops. You must also be wary of them becoming frozen into the ground, making it impossible to dig them up.

Planning your harvest

In order to avoid having to harvest all your produce at the same time, try to plan what you are going to grow and when, to make the best of the growing season. Follow the chart on the next page for the sowing (S) and harvesting (H) times of the most popular vegetables, but refer to seed packets for specific instructions for your chosen variety.

Crop	Jan	Feb	Mar	April	May	June	July	Aug	Sept	Oct	Nov	Dec
Artichokes, Jerusalem	H	S	S							H	H	H
Beans, dwarf French				S	S	H	H	H	H	H		
Beans, runner					S	S	H	H	H	H		
Beetroot				S	S	S	S/H	H	H	H		
Broad beans		S	S	S			H	H		S	S	
Brussels sprouts	H	H	H	S/H	S				S/H	H	H	H
Cabbages			S	S	S	H	S/H	S/H	H	H		
Carrots			S	S/H	S/H	S/H	S/H	H	H	H		
Courgettes and marrows					S	S	H	H	H	H		
Cauliflowers			S	S	S	S	H	H	H	H	H	
Celery					S	S			H	H	H	
Cucumbers					S	S/H	H	H				
Garlic (cloves)	S						H	H	H	S	S	S
Leeks	H	H	S	S						H	H	H
Onions (sets)			S	S				H	H	H		
Parsnips	H	H		S	S					H	H	H
Peas			S	S	S	S/H	S/H	H	H	H		
Potatoes			S	S	S		H	H	H	H		
Shallots		S	S	S			H	H	H			
Squashes and pumpkins			S	S	S			H	H	H		
Sweetcorn			S	S	S		H	H	H			
Swiss chard			S	S	S/H	S/H	S/H	H	H	H		
Swedes					S	S	S			H	H	H
Tomatoes		S	S			H	H	H	H			
Turnips			S	S	S/H	S/H	S/H	H	H	H		

Crop rotation

To keep digging and manuring to a minimum, whatever size your plot, divide it into four sections and rotate your crops each year, in order to make the best use of the soil. If you manure it well the first year before planting cabbage and again the next year before planting potatoes, you will only need to give the soil a light top dressing of manure in the following years.

Year	Section A	Section B	Section C	Section D
1	Greens	Peas and beans	Roots	Potatoes
2	Peas and beans	Potatoes	Greens	Roots
3	Potatoes	Roots	Peas and beans	Greens
4	Roots	Greens	Potatoes	Peas and beans

2 Gathering in the harvest

So your crops are ripening fast and your fruit trees are weighed down with fruit (or there is a good pick-your-own farm nearby). You now have more fresh food than you can eat but you don't want it to go to waste. Now is the time to store it for later when the season is over. Here are a few basic rules for getting the best out of your produce:

o   Decide first where and how you are going to store your produce, and prepare the space and equipment necessary.
o   Harvest produce when it is in peak condition. If it is past its best, flavour will be lost, and it will rot quickly if in dry storage.
o   When picking, handle carefully to avoid bruising.
o   Harvest only when you are ready to prepare the produce for storage, whether it be in the freezer, as conserves, wine, in sand boxes or strung up for drying.
o   Process as quickly as possible after harvesting for the best quality.
o   Label everything with a description and date.
o   Keep an eye on produce that is being stored dry in a shed or in boxes. Remove any that show signs of deterioration to prevent them from contaminating their neighbours.

3 Storage

If you have a shed or outbuildings, a cellar or some space that isn't heated during the winter, you are in the lucky position of being able to grow many varieties of crop with no concerns of where to store them if they ripen at the same time.

There are several methods of simple dry storage:

o Packing in sacks or boxes and keeping in the dark, e.g. as for potatoes
o Layering in boxes with sand, straw or dry soil, e.g. carrots and beetroot
o Wrapping individually in newspaper and storing side by side in boxes or on trays, e.g. apples
o Drying and hanging in a cool, dry, well-ventilated place, e.g. onions
o Leaving in the ground throughout winter, e.g. leeks and parsnips.

Remember to ensure that the produce you harvest is in peak condition. It must be dry and free from any bruising or damage that will cause it to rot and spread to the rest. There is no point in storing a fruit or vegetable that has the smallest bruise or pest hole in it, as it will only get worse.

Where to store

o Farm outbuildings
o Garage – if it has rafters, planks can be rested across them to provide a flat surface
o Loft, if it fits the criteria below
o Cellar that is dark, moist but well ventilated
o Under the bed in a spare, unheated bedroom
o Shed

However, as different types of produce need different conditions, it is important to keep to the following criteria when choosing a storage place to protect your harvest:

o **Frost free:** If you only have a shed, you may have to give added protection to your produce in a harsh winter by covering it with horticultural fleece or bubble wrap.
o **Bug, slug and rodent free:** Inspect your chosen storage area thoroughly for any signs of previous pest visitations and deal with accordingly to protect your harvest.
o **No strong-smelling products:** They will taint produce in storage nearby. In addition, fruits and vegetables may taint each other, so avoid storing onions or garlic in the same area as grapes, or pears and apples together, notwithstanding that they need to be stored at different temperatures.
o **Good ventilation:** This is important to prevent rot, but the air should be a little moist, especially for roots and apples.

Storage equipment

You do not have to spend a lot of money buying boxes and bags for storage. You will be surprised what you might have squirreled away in a shed or the attic that could be useful. Old net curtains, for example, could be used for sacks or for hanging up onions; use old furniture such as a cupboard or sets of drawers, or a defunct fridge or freezer in the garage or shed. Just make sure that you make ventilation holes that don't admit mice.

Basically, you need any of the following:

o   Well-ventilated plywood boxes, such as those in which a greengrocer or supermarket receives fruit and veg deliveries
o   Plastic stacking boxes – the holey variety
o   Plastic netting, or net or cotton shopping bags for hanging up onions or pumpkins
o   Hessian or sturdy paper agricultural sacks (which must not be allowed to sit on a damp surface), or plastic netting ones
o   Large trays for apples or pears, or sheets of hardboard or plywood.

What to store

Storage method	Crop
In the ground	Brussels sprouts, cabbages, carrots, celeriac, celery, chicory, Jerusalem artichokes, kale, leeks, parsnips, salsify, spinach, sprouting broccoli, swedes, Swiss chard, turnips
In a clamp (page 18)	Beetroot, carrots, celeriac, kohlrabi, parsnips, potatoes, salsify, swedes, turnips
Layered in boxes	Beetroot, carrots, celeriac, kohlrabi, leeks, parsnips, salsify, swedes, turnips
In bags	Kohlrabi, potatoes, swedes, turnips
On trays	Apples, pears, quinces, squashes
Hung up	Garlic, onions, shallots, squashes, marrows, pumpkins

How to store

Bagging up

This is the simplest, most trouble-free method of storage. With root vegetables such as potatoes and turnips, harvest on a dry and sunny day if possible and allow them to dry enough so that you can rub off some of the soil. There is no need to wash them. Remove any that are damaged or look as though they might have slugs – a tiny hole can reveal a network of tunnels in a potato leading to a tiny slug – then place the rest in a large agricultural paper bag or, better still, a hessian sack and store in a cool, dark place that is slightly humid. Check regularly for bad ones and make sure the store is rodent proof.

Layering in boxes

Root vegetables such as carrots, parsnips and beetroot can be stored layered in boxes of moist (but not wet) sand, soil, peat, sifted wood ash or even shredded paper, which can be composted afterwards. Use wooden greengrocer's boxes or plastic basket-style boxes that allow the air through. Allow the roots to dry and brush off excess soil, cut off the leafy tops close to the crown. Twist the tops off beetroot to prevent the red juice bleeding.

Put a layer of the sand, ash, soil or peat in a box and cover this with the roots, then cover with another layer of sand, etc., then another layer of roots, repeating until the box is full. Keep in a cool shed or garage in the dark and use when needed.

On trays

Ideally, easily damaged fruit such as apples and pears should be carefully laid out on trays lined with kitchen towel, newspaper or shredded paper to make a soft bed. The fruit must be absolutely perfect with their stalks still on and they must not touch each other, in case one deteriorates and contaminates the rest. The trays should then be placed in a cool but frost-free, moist but well-ventilated space. If you do not have the room, you can wrap them individually in newspaper and store them in boxes insulated with bubble wrap.

The fruit must be handled very carefully to avoid bruising. Do not store any that are damaged as they will rot. Pears need slightly warmer and more humid conditions than apples and can be wrapped very loosely.

Hanging up

Vegetables that need to dry out a little, such as onions and squashes, store best in net bags hung in a cool, dry area or in large nets slung between the rafters of a garage or shed, where the air can circulate well. Onions, garlic and shallots are harvested when their leaves have dried and died down so they can be plaited together in an attractive rope and hung up to dry, but they must be completely dry first as any moisture in the plait will cause them to go mouldy. They should not be stored in an area with any other fruit or veg because of their strong smell. The same goes for cabbages, which should be hung upside down by their roots. Squashes need a warmer storage area, such as a room in the house.

In the ground

Some crops, such as cabbages, leeks, parsnips, swedes and celeriac, can be left in the ground over winter and harvested when needed. In a harsh winter when the ground is permanently frozen, they will be difficult to extract from the soil, so if a hard, cold spell is forecast, dig out a few weeks' worth and store layered in boxes, as above. Alternatively, cover the crops with a 30 cm (12 in) layer or more of leaves, straw or hay, which will also prevent them from becoming saturated in wet weather.

In a particularly wet winter, crops would also be better stored layered in boxes.

Storing root vegetables in a clamp

Up until the middle of the last century, before supermarkets provided all our culinary needs, when families had to be self sufficient to survive, root vegetables were kept fresh throughout winter by storing them outside in a clamp. If you have the space in your plot or allotment, you can still do this today.

1   Dig 2 parallel trenches, about 10–15 cm (4–6 in) deep, and 1 m (3 ft) apart, for drainage, heaping the soil in between to make a raised area for the vegetables, which keeps them away from ground water and frost.

2   Spread a layer of straw over the raised level and pile the root vegetables in dry layers on top to form a pyramid.

3   Cover the heap completely with straw and then with another layer of soil so that it looks like a 1 m (3 ft) high molehill. The soil causes the straw to decompose and create enough warmth within the clamp to prevent the vegetables from freezing, while at the same time keeping out the rain.

4   Dig the roots out from the top when needed, and replace the straw and soil carefully.

For a more modern version:

1   Spread plastic sheeting on some spare ground.

2   Cover thickly with straw and layer the vegetables on top, interspersed with dry sand.

3   Cover completely with more straw and a layer of soil, and finish with another sheet of plastic, weighted down around the sides to prevent it from being blown off. You could also use old blankets, carpet, newspaper or shredded paper.

4 Drying

One of the first methods of preserving food that dates back to ancient times, drying is simply a process of removing all the moisture from produce by a low heat to prevent the growth of bacteria, yeasts and destructive enzymes. Food was first dried in hot countries by laying it out in the sun; where long enough periods of sunshine cannot be guaranteed, warm, well-ventilated places must be found instead.

Drying food concentrates the flavour and retains the vitamins, minerals and other nutrients and is a particularly good way of storing herbs. Dried produce is easy to store in bags or jars in a cupboard, but for optimum quality is best kept for no longer than a year. It is also convenient and simple to use, either by soaking first to rehydrate or just adding dried vegetables straight into stews and soups, and fruit into cereals and baking, or eating as snacks.

What and how to dry

Practically everything is easy to dry but not everything is worth it, as the flavour and colour may not be as appetising as other methods of preservation. Peas, for example, dry very well, but retain more goodness, colour and flavour when they are frozen, and root vegetables dry well without too much hard work. The following are worth drying:

o Apples
o Grapes
o Herbs
o Pears
o Peppers
o Plums
o Pumpkin seeds
o Tomatoes

The main aim is to get as much dry air freely circulating around the produce as possible. It doesn't have to be warm air, but the process will take longer if it is cold. If the air is too warm, however, the skins of the produce will become tough. The ideal temperature for drying is 50–60°C/120–140°F.

Equipment

o A warm airy place such as an airing cupboard, boiler room
o Warming oven of an Aga or Rayburn, or an electric fan oven set on low with the door open
o Wire or wooden racks, such as a cake cooling rack
o String and bamboo poles for threading fruit or vegetable rings on, or for tying up herbs, which can also be attached to lines with wooden pegs

If you want to go into drying big time, or to make sure you get it absolutely right, you could invest in a fruit and vegetable dryer or dehydrator, but they are expensive. Alternatively, you could make your own solar dryer (page 24).

How to dry herbs

Preparation
Not all herbs dry well, but those with a low moisture content are suitable, e.g. bay leaves, thyme, rosemary, oregano, mint and sage. For optimum flavour, harvest the herbs just before they have reached flowering stage on a warm dry morning, after the dew has dried on them and before the sun has warmed them too much, causing their essential oils to evaporate. Remove any damaged or diseased leaves and shake out any insects. If necessary, wash in cool water and pat dry with kitchen paper.

Air drying
This is the best method of drying herbs as their natural oils are preserved, creating a better flavour. Make sure the herbs are free of moisture before removing the leaves from the last 2–3 cm (1 in) of stalk and tying in small bunches. Each variety should be dried separately. Hang upside down or peg the sprigs to a line and leave in a warm, airy, dust-free place out of direct sunlight, which would affect the volatile oil content and lessen the scent and flavour.

Be careful not to make the bunches too large as the herbs in the centre will become moist and go mouldy. Label each herb, as it is sometimes difficult to tell which is which once they have dried.

Oven
If you have no space to air dry herbs, dehydrating them in the oven is second best. Make sure the oven is cool (40–50°C/100–120°F) and if it is a fan oven all the better. Herbs exposed to too much heat will turn brown and lose flavour.

Remove the leaves from larger leaved herbs such as sage and mint, but leave the smaller leaved variety on their stems. Ensure they are free of moisture, spread on a wire rack and leave in the oven for about an hour until crisp.

Microwave
A microwave set on defrost can be used to dry herbs, but applying microwave heat literally cooks the herbs robbing them of their volatile oils.

Drying time

If the herbs are drying in a cool outbuilding, they will take 3–4 weeks; in an airing cupboard, 3–4 days; above a range oven, 3–4 hours; in an oven, 1 hour. The herbs are ready when the stalks can be snapped off and the leaves rubbed off their stems.

Storage

If possible, keep the leaves whole until you are ready to use them in order to preserve their flavour. Discard any that are brown or mouldy. Gently rub the leaves off their stems with your fingers or by rolling lightly with a rolling pin. Store them in well-labelled, airtight, dark glass jars in a cool, dark place and use within a year.

To use, sprinkle straight into dishes when cooking.

How to dry seeds

To dry seeds such as dill, fennel and coriander:

1 Gather the herbs on a warm, dry day after they have flowered when the seed capsules have turned brown.
2 Cut off the whole seedheads about halfway down the stem, tie them in a bunch and hang them upside down with their heads in a paper bag, allowing plenty of air inside.
3 Hang in a cool, dry place until all the seeds have fallen off.
4 Spread the seeds on trays and leave in an airy room covered with a thick layer of newspaper for two weeks.
5 The seeds are ready to be put into clean dry jars when they are hard and difficult to break with a fingernail. Do not store them if they smell musty.

How to dry flowers

To dry flowers such as marigolds, nasturtiums and borage for salads and garnishes:

1 Harvest flowers when they are fully open and in good condition. They must not be wet.
2 Handle them gently and do not wash them. Spread them out on a muslin-covered rack making sure that air can circulate freely around them. The petals are fleshy and will give off a lot of moisture.
3 Leave in an airing cupboard or similar warm dark dry place until papery dry.

How to dry fruit

Preparation

Fruit to be dried must be fresh, ripe and in pristine condition, as well as clean and dry. Thick-skinned fruit such as apples and pears should first be peeled, cored and sliced (halved or quartered for pears) and, if they turn brown easily, plunged into acidulated water (5 ml/1 tsp lemon juice to 300 ml/½ pint water, or 50 g/2 oz salt to 4.5 litres/1 gallon water). Keep the slices thin (about 5 mm/¼ in), but not too thin or they will become crispy; too thick and they'll take longer to dry and eventually go mouldy. Try to slice the fruit evenly, so the pieces will dry at the same time.

Remove the stones from fruit such as plums and damsons and dry them whole, or in halves for larger fruit such as peaches. Tomatoes can be halved depending on their size, but the seeds don't have to be removed.

Air drying

Cut the fruit into rings and slide them on to horizontal bamboo poles or string pulled taut when hung. Do not allow the rings to touch each other, for maximum air circulation. Swap them around if they are drying unevenly.

Oven

Lay sliced or whole fruit, depending on the type, in single layers on wire or wooden racks. Line them with muslin if the fruits, such as grapes, are likely to fall through the gaps once dried. If using a gas oven, ensure that the material does not dangle in the flame.

Place the trays in a cool fan oven with the door ajar and heat up to no more than 70°C/150°F/gas 3 very slowly. If the fruit has been strung up and is taking too long to dry, you can finish them off in the oven.

Microwave

Space the prepared fruit out evenly on baking paper on the turntable and set the oven to defrost. Microwave for 5–10 minutes at a time, turning the fruit occasionally until it is dehydrated. Remove from the microwave and place on a wire rack to complete the process.

Drying time

The fruit should warm up slowly to 50–60°C/120–150°F to prevent the skin from hardening. Once the surfaces have formed a seal, the insides will dry more slowly depending on the thickness of the pieces. Overall drying time can take from a few hours to several days or weeks depending on whether you're using an oven, microwave or outbuilding. To test that the fruit is completely dry, squeeze it gently with your fingers and if no moisture comes out and it springs back into shape, it is ready. The fruit should be soft and flexible but not juicy; any moisture left inside will encourage the growth of harmful bacteria.

Storing

Ensure the fruit is completely cool before packing into sterilised airtight containers.

To use

Either eat straight from the jar without rehydrating, use directly in baking (buns, cakes, biscuits) or soak first in cold water overnight and add to winter fruit salads, or cook after soaking and use in desserts or tagines, curries and stews.

How to make your own solar dryer

You need a metal cake rack; a sturdy cardboard box 5 cm/2 in bigger than the rack all round; a small bowl, half the depth of the box; silver foil; fine insect-proof netting; scissors.

1 Cut away the top of the box and half of one side.

2 Line the box with the foil, shiny side up to reflect the sunlight.

3 Place the bowl bottom side up in the centre of the box and rest the rack on top to allow plenty of air circulation.

4 Lay the prepared produce on the rack, well spaced.

5 Cover the box tightly with the netting and place in the sun.

Tomatoes, plums and grapes dry well using this method, as do peppers. It may take two days or a week depending on the weather – the intensity of the sun and the dryness of the air. Take the box in at night and when it rains.

How to dry vegetables

Preparation

Vegetables are dried in the same way as fruits and the same rules apply. Only use the most perfect fresh specimens, which should be washed and dried carefully, or wiped with damp kitchen paper. Peel vegetables with tough skin, such as root vegetables. Peppers should be fully ripe and red before harvesting with some stalk remaining.

Leave peas and beans on the plants until the pods have dried and shrivelled, then cut the plant at the base and hang indoors to complete the drying process. Remove the seeds from the pods when hard and store in airtight containers.

Air drying

A needle and heavy-duty thread will come in handy for air-drying root veg and peppers. The latter – both sweet and chilli – look wonderful being air dried. String them together by sewing the peppers to one another through the tops of the fruit, under the stalk, to form a ristra, as the South Americans and Basques do. Hang them in a warm, dry place – ideally in a greenhouse – and on sunny days hang outside in the sun to intensify the flavour.

Root vegetables can be sliced thinly lengthwise and threaded with a needle, leaving space between them, before being hung up in a dry, airy place, but it is very labour intensive and many people feel it is not worth the bother.

Oven

Slice prepared vegetables according to their type and lay out on racks lined with muslin in a cool oven between 50–60°C/120–150°F. Do not slice peppers too thinly. Halve or quarter them and remove the seeds.

Drying time

The fleshier peppers will take a few weeks to dry naturally but just a few hours in the oven; root vegetables will take a week or two naturally.

Storage

Dried strings or ristras of peppers can remain as they are, enhancing the kitchen decor, but other dried vegetables should be kept in clean, airtight containers.

To use

Dried vegetables can usually be added straight into stews without rehydrating first, if the stew is going to be cooked slowly.

5 Freezing

Most people who grow their own food find that freezing their harvest is the most convenient way of storing it for year-round consumption. It is simple and quick to do and you don't need masses of space, unless you have a field of produce to preserve that is ready to harvest all at once. The food is also quick to access as it can usually be cooked straight from the freezer.

Freezing food to preserve it was practised long before electricity was discovered. In prehistoric times, meat was packed in snow and ice to make it last longer, and in the Middle Ages cellars were filled with ice in winter in the hope that food stored there would last during the summer. In the late 19th century, mechanical freezing methods were developed and today few of us would be without a freezer.

Speed is very important in the freezing process and produce must be frozen as soon after harvesting as possible in order to halt deterioration and to preserve the vitamins and nutrients. Once frozen, the micro-organisms that have started the rotting process are stopped but not killed, so as soon as a food has thawed it will start to deteriorate again.

What to freeze

Most fruits and vegetables are suitable for freezing, but some produce with a high water content, such as strawberries, tomatoes and salads, undergo a structural change when frozen and are usually better cooked before freezing. Root vegetables are not worth freezing unless they are young and cannot be dry stored.

Equipment

Freezer
Chest freezers, even the small ones, are more capacious than upright ones with drawers. You can fit big awkwardly shaped bags of dry frozen produce in them that are easily accessed with a scoop. However, an upright is more convenient for smaller quantities because it takes up less space and can be part of a fridge-freezer in the kitchen.

It is important you maintain your freezer well, keeping it clean and frost free. Some freezers are self-defrosting but if not, you will need to defrost a chest freezer at least once a year and two or three times for an upright model. In between, scrape off any accumulated frost with a plastic scraper.

Freezer thermometer
This is helpful but not necessary. It is useful to check the temperature inside the freezer from time to time. It should be kept at -18°C/0°F. For fast-freezing large quantities of produce, the temperature should drop to -25 to -30°C/-13 to -22F.

Polythene bags
For freezing, you need heavy duty plastic bags that do not puncture easily. It is best to buy those sold as freezer bags in different sizes, along with ties with which to seal them if they do not have a zip. All the air must be removed from a bag before sealing it tightly and this can be done by inserting a drinking straw into the opening and sucking out the air.

Rigid containers
An assortment of rigid, plastic boxes with airtight lids is useful for dry-packing fruit with sugar or freezing it in syrup, or freezing any other liquid-based or semi-solid food.

Foil containers
Foil containers with fitted lids are ideal for cooked dishes as you can put them straight into the oven from the freezer to heat up.

Labels
Labelling is very important, of course. Every item must be marked with what it is and the date it was frozen. It is also useful to write

down the weight or quantity. If it is a cooked dish, mark on how many it will feed. Buy labels that are specifically for freezing so that they stick properly and remain on the bag without falling off over time, but make sure you stick them on to a clean dry surface to begin with. Alternatively, you can buy a freezer pen that will write on bags and plastic boxes indelibly.

Blanching basket

A blanching basket or chip basket (line with muslin cloth for peas or other smaller items) or a collapsible vegetable steamer is advisable for blanching, as you are able to plunge the vegetables into boiling water all at once and remove them to cool down with optimum speed.

Freezing tips

1   Do not harvest your produce until you are ready to begin preparing it for the freezer, ideally not much longer than two hours ahead.

2   Process only small quantities at a time, so that you can get at least some produce into the freezer before possible interruption.

3   Complete one batch at a time. Do not have several batches on the go at different stages.

4   If you have a lot to pick and it must be done in one go, wrap what you pick in newspaper and put it in the shade to keep cool while you continue harvesting.

5   Keep the produce in the fridge if you cannot freeze it straight away.

6   Organise your freezer well. If you have a chest freezer put new batches of food at the bottom so that the older produce gets used first. If you have an upright, keep fruit in one drawer, vegetables in another, and strong-smelling vegetables such as onions in a separate one, double wrapped to prevent tainting.

7   Keep a diary recording when you put produce in the freezer, the date it should be eaten by and, if you have a chest freezer, where it is. Cross it out when you have removed it.

How to freeze

Although most fruit and vegetables freeze well, they do not all freeze in the same way. As with most storage methods, the rule is that only top-quality produce should be frozen, with the exception of apples, which can have the bad bits cut out, as they will be cooked after freezing. There are several methods:

o Open freezing
o Blanching
o Dry-packing
o In syrup
o Purée
o Cooked

Open freezing

Many fruits freeze well by just being cleaned, preferably not washed (but if it is necessary, wash very gently in ice-cold water and pat dry with kitchen towel), picked over and hulled, then laid out on lightly oiled baking trays and put into the freezer. When frozen you can pour them into freezer bags, remove the air with a straw, seal and return to the freezer.

Apples, blackberries, blueberries, currants (black, white and red), gooseberries, raspberries, loganberries are suitable.

Blanching

It is essential for vegetables to be blanched (scalded) before freezing to destroy the enzymes that cause them to deteriorate, and to help maintain their colour, texture, flavour and vitamin content.

1 Prepare the vegetables appropriately.
2 Bring a large pan of water to the boil, place 450 g/1 lb vegetables in a blanching basket and plunge into the boiling water.
3 Bring back to the boil as quickly as possible and boil for the recommended time, often only 1 minute.
4 Drain immediately and plunge the vegetables into a bowl of iced water to cool them down as quickly as possible; or hold them under cold running water for the same amount of time it took to blanch them.

5   As soon as they are cool, drain and pat dry with kitchen paper. Pack in freezer bags in convenient quantities, remove the air, seal, label and freeze. Blanched vegetables can also be open frozen.

Dry-packing

This is another quick way of freezing fruit and vegetables without adding liquid. Prepare your produce according to type, then pack in rigid containers, leaving 2.5 cm/1 in head space. Fruits, such as halved and stoned plums, can be layered with sugar: 100–175 g/4–6 oz to each 450 g/1 lb fruit.

Vegetables should be blanched first but make sure they are completely dry to prevent them from freezing together in a lump. Remove the air by lifting the closed lid at a corner and pressing on the box to squeeze it out. Quickly reseal.

In syrup

Freeze fruits in syrup and you can use the fruit-infused syrup when thawed as well. The strength of the syrup needed depends on the acidity of the fruit, but most just need a light syrup made with 225 g/8 oz granulated sugar for every 600 ml/1 pint water. A medium syrup is made with 225–350 g/8–12 oz sugar and a heavy syrup with 350–450 g/12 oz–1 lb to 600 ml/1pint water.

Boil the sugar and water together until dissolved and continue boiling for 2 minutes. Cool before using and make sure it is completely cold before putting in the freezer. You will need enough to cover the fruit completely in rigid containers, leaving 2 cm/¾ in headspace. Screw up balls of greaseproof paper and place them on top of the syrup to keep the fruit from floating to the top. Some fruits, such as figs, may need to be poached in the syrup for 4–5 minutes first.

Purée

Soft fruits, such as strawberries, can be puréed raw with or without sugar and placed into small rigid boxes for use in sauces and coulis, etc. Stone fruit should be poached gently first in a small amount of water and sugar according to how sweet the fruit is and allowed to go cold before freezing.

Best freezing methods

Fruit or vegetable	Blanching time (minutes)	Method	Storage time (months)
Apples, peeled, sliced, quartered	–	Open freeze, pack in freezer bags	12
cooked purée	–	Pack in freezer bags	8
Apricots, peeled halved	–	Medium syrup, pack in rigid containers	12
Asparagus, sprue	1		
medium stalks	2	Pack in rigid containers	1
thick stalks	3		
Beans, French, whole/ large pieces	2–3	Pack in freezer bags/rigid containers	12
Beans, runner, chopped	2–3	Pack in freezer bags/rigid containers	12
Beetroot, cooked, chunks	–	Pack in freezer bags/rigid containers	12
Blueberries, whole	–	Open freeze, pack dry medium-heavy syrup, pack in freezer bags/ rigid containers	12
Blackberries, whole	–	Open freeze, pack dry medium-heavy syrup, pack in freezer bags/ rigid containers	12
Broad beans	2–3	Open freeze, pack in freezer bags/rigid containers	12
Broccoli, spears	2–4	Open freeze, pack in freezer bags	12
Brussels sprouts, whole	2–4	Open freeze, pack in freezer bags	12
Cabbage, green/white, shredded	1	Pack in freezer bags	6
red, shredded	1	Pack in freezer bags	12
Carrots, baby new	3	Pack in freezer bags/ rigid containers	12

Fruit or vegetable	Blanching time (minutes)	Method	Storage time (months)
Cauliflower, florets	3	Open freeze, pack in freezer bags	6
Celeriac, diced;	1	Pack in freezer bags	6
grated	2	Pack in freezer bags	6
Celery, chunks	3	Pack in freezer bags	6
Cherries, stoned	–	Medium–heavy syrup, pack in rigid containers	12
Chillies, split, deseeded	1	Pack in freezer bags	6
Courgettes, large, sliced	1	Open freeze, pack in freezer bags	12
small whole		Open freeze, pack in freezer bags	12
Currants, black/red/white, whole	–	Open freeze, pack dry, pack in freezer bags/ rigid containers	12
crushed/uncooked purée	–	Pack in freezer bags/ rigid containers	12
Damsons, whole;	–	Open freeze, pack dry, pack in freezer bags/ rigid containers	6
halved, stoned;	–	Pack dry/medium syrup, in rigid containers	12
cooked purée	–	Pack dry/medium syrup, in rigid containers Pack in freezer bags/ rigid containers	6
Fennel bulb, quartered	3–5	Pack in freezer bags	6
Figs, fresh, whole, unpeeled	–	Open freeze, pack dry, in freezer bags/ rigid containers	12
whole, peeled/sliced	–	Light–medium syrup, pack in rigid containers	12
Gooseberries, whole	–	Open freeze, pack dry, in freezer bags/ rigid containers	12
cooked purée	–	Pack in freezer bags/ rigid containers	12

Fruit or vegetable	Blanching time (minutes)	Method	Storage time (months)
Grapes, seedless, whole, in bunches	–	Open freeze, pack dry, in freezer bags/ rigid containers	12
Large, halved, seeded	–	Light syrup, pack in rigid containers	12
Kale (curly), leaves	1	Pack in freezer bags	12
Kohlrabi, large, sliced/diced	2	Pack in freezer bags	12
small, whole	3	Pack in freezer bags	12
Leeks, large, sliced	3	Pack in freezer bags	6
small, whole	4	Pack in freezer bags	6
Loganberries, whole	–	Open freeze, pack dry/ medium syrup, pack in freezer bags/rigid containers	12
crushed/cooked purée	–	Pack in freezer bags/ rigid containers	12
Mangetout, whole	½–1	Pack in freezer bags	6
Marrow, chunks	2	Pack in rigid containers	6
Nectarines, peeled, stoned, halved/sliced	–	Heavy syrup, pack in rigid containers	12
Onions, chopped	1	Double wrap and pack in freezer bags	3
button/pickling	2	Open freeze, double wrap and pack in freezer bags	3
Parsnips, sliced	2	Pack in freezer bags	6
Peaches, peeled, stoned, halved/sliced	–	Heavy syrup, pack in rigid containers	12
Pears, firm eating/cooking, peeled, halved, cored, quartered/sliced	–	Heavy syrup, pack in rigid containers	12
Peas, podded	1	Open freeze, pack in freezer bags	12
Peppers, chopped/sliced	2	Open freeze, pack in freezer bags	6
whole, deseeded	3–4	Open freeze, pack in freezer bags	6

Fruit or vegetable	Blanching time (minutes)	Method	Storage time (months)
Plums, halved, stoned	–	Medium–heavy syrup/pack dry, in freezer bags/rigid containers	12
Quinces, peeled, cored, sliced, poached	–	Pack in rigid containers	12
Raspberries, whole	–	Open freeze, pack dry in freezer bags/rigid containers	12
uncooked purée	–	Pack in freezer bags/rigid containers	12
Rhubarb, chunks	–	Open freeze, pack dry in freezer bags/rigid containers	12
Salsify/scorzonera, cut into short lengths after blanching	2	Pack in freezer bags	12
Shallots, whole	2	Open freeze, double wrap and pack in freezer bags	3
Spinach, leaves	2	Pack in freezer bags	12
cooked purée	–	Pack in freezer bags	12
Squashes and pumpkins, diced	1	Pack in freezer bags	12
Strawberries, uncooked purée	–	Pack in rigid containers	12
Swede, diced	2	Pack in freezer bags	12
Sweetcorn, cobs	2–8	Open freeze, pack in freezer bags	12
kernels	–	Pack in freezer bags	12
Swiss chard, leaves	2	Pack in freezer bags	12
ribs/stalks	3	Pack in freezer bags	12
Turnips, large, sliced	2	Pack in freezer bags	12
small whole	4	Pack in freezer bags	12

Cooking for the freezer

Some fruit and vegetables are not worth freezing raw as it changes the consistency and taste too much, so it is best to cook them before freezing. This way you have portions of fruit and vegetables that just need heating up once thawed. Some you can heat from frozen.

Alternatively, you can cook some of your produce into purées, sauces, soups, bakes and gratin dishes or stew fruit to keep in the freezer to make life easier later on in the year. The advantage of this is that you can use produce that is not top quality and it isn't so important to get it from garden to freezer in ultra-quick time.

What to cook and freeze
o Apples
o Beetroot
o Damsons
o Gooseberries
o Plums
o Pumpkin
o Quinces
o Squashes
o Tomatoes

To freeze cooked dishes
o **Soups:** Pour cold soup into a freezer bag and freeze. Reheat slowly from frozen.
o **Bakes and gratins:** Line a serving dish with foil and fill with the cooked vegetables and sauce and sprinkle with grated cheese and breadcrumbs (if using). Open freeze until firm, remove the dish and wrap the frozen food well with more foil, seal in a freezer bag with air removed, label and return to the freezer. To defrost, remove from the wrappers, thaw at room temperature for 3–4 hours, then bake in an oven preheated to 190°C/375°F/gas 5 for 35–40 minutes until golden and heated through.

Thawing and using

How you treat your produce when you take it out of the freezer is as important as how you grow and freeze it, to ensure the food looks and tastes its best. Once the food has been defrosted, eat or cook immediately.

To eat raw

You will want to maintain the shape and firm texture of the fruit or vegetable that you plan to eat raw, so it must be thawed slowly to the point where it remains chilled. Rapid thawing or thawing for too long will soften it and risk it going mushy, so it is best to place the bag of frozen fruit or vegetables in the fridge – a 450 g/1 lb bag will take 8–10 hours to defrost in the fridge.

For food that has been open frozen, tip out the quantity needed into a wide bowl and place in the fridge; because the produce is not clumped together, this will not take long to defrost – 3–4 hours at most. If time is short, defrosting at room temperature halves thawing times, and for a speedier option, run the pack under a cold tap, but the texture and flavour will suffer. Alternatively, defrost in a microwave, following the manufacturer's instructions.

Once food has been defrosted, it is not safe to refreeze.

For cooking

Frozen produce to be recooked is best thawed quickly to ensure it is soft and tender. If fruit is to go into a tart, it must be completely thawed to prevent the pastry from going soggy; into a pie without a pastry base, fruit can be partially thawed, enough to be able to break it up and fit it into the dish. The juices that defrost during cooking will enhance the dish.

Fruit or vegetables that have been open frozen can be added to a dish without being defrosted first, as long as the fact there will be some extra liquid created during cooking is taken into consideration.

Fruits packed in syrup need more thawing time than those packed in sugar.

Cooking from frozen

Vegetables can be cooked from frozen by just plunging them into boiling salted water and cooking until tender. Remember, though, that if they have been blanched already, cooking time will be less.

6 Bottling

Bottling is another traditional way of preserving your precious produce, but it is only fruit (including tomatoes) and a handful of vegetables that can be bottled at home. Most vegetables do not have enough acid to prevent the growth of the micro-organisms that cause botulism and it is not usually possible to obtain the temperatures necessary to guarantee complete sterilisation.

Successful bottling depends on heating the fruit in water or syrup in jars until sterilised. An airtight seal is then made to prevent the infiltration of any further micro-organisms that could contaminate the contents. Another way of preserving your fruit is with alcohol and sugar. These ingredients are preservatives in their own right so no heating is required and the end result makes a very fine dessert or digestif.

Vegetables and tomatoes need to be soaked in brine first.

What to bottle

o Apricots
o Artichoke hearts
o Asparagus
o Blueberries
o Blackberries
o Cherries
o Currants
o Damsons

o   Gooseberries
o   Peaches and nectarines
o   Pears
o   Plums and greengages
o   Raspberries and loganberries
o   Tomatoes

Equipment

Glass preserving jars

Always use clean, sterilised jars and lids without ridges or chips where bacteria can lodge, and be sure that the rubber bands are soft and flexible if they are not attached to the lids.

o   Glass jars such as Kilner jars are ideal, as they have a rubber band around the neck on which the glass tops are clipped to create a vacuum during cooling. When the jars are cold, the vacuum formed inside holds the lid in place. Metal clips seal the tops down and also prop the lid up during the bottling process.

o   Jam jars with special metal tops kept in place with a plastic or metal screw band can also be used. During bottling, the screw band is not tightly secured to prevent the bottle from exploding.

Pressure cooker

A pressure cooker makes the whole bottling process much simpler and quicker. Use the trivet for the jars to sit on so that the glass does not come into contact with the metal cooker or each other.

Preparation

Fruit and vegetables should be carefully prepared for bottling and should be in perfect condition and ripe and firm but not over-ripe. Remove stalks and leaves, rinse in ice-cold water and pat dry. Large fruit may be halved and stoned. Fruit with light-coloured flesh, such as pears, should be covered with syrup or acidulated water (5 ml/1 tsp lemon juice to 300 ml/½ pint water, or 50 g/2 oz salt to 4.5 litres/1 gallon water) as soon as possible to prevent browning.

Fruits can be bottled in plain water, syrup or dry-packed with sugar. Syrup maintains a better flavour and colour and also means that fruit is ready to use straight from the jar. Asparagus, tomatoes and artichoke hearts are best bottled in brine.

Syrup

A ratio of 450 g/1 lb granulated sugar to 1 litre/1¾ pint water gives a suitable strength of syrup for most fruit. The sugar and water should be boiled for 1 minute before use (either hot or cold, according to the method). Add any flavourings, such as spices or orange, at this stage.

Brine

Dissolve 15 ml/1 tbsp fine salt in 1 litre/1¾ pint water. You can also add 10 ml/2 tsp lemon juice and 2.5ml/½ tsp sugar to each 450 g/1 lb jar to improve the colour and flavour of asparagus, tomatoes and artichokes.

How to bottle

1   Pack fruit or vegetables tightly a third at a time into sterilised jars standing in a bowl half-full of boiling water to keep the temperature from dropping too much (except for the slow water bath method, page 42).
2   Gradually pour over the boiling syrup, brine or water to just cover the produce as each third goes in. Twist the jars from side to side to get rid of any air bubbles. Space must be left at the top to prevent any liquid from boiling over.
3   If you are dry-packing fruit in granulated sugar, sprinkle the sugar over each layer as you pack the jar.
4   Wipe the necks of the jars, dip the rubber bands in boiling water to soften them, and screw or half clip on the tops. With a screw band top, tighten it securely then unscrew it by a quarter turn.
5   After they have been processed, tighten the screw bands or push down the clips and leave the jars to cool until they are completely cold.
6   Remove the screw bands or undo the clips. Carefully lift the jars by their lids to test that a vacuum has been created and they remain firmly sealed. If a lid comes off, use the contents immediately and check the jar, lid, screw band or clip for a fault.
7   Label and store in a cool, dry place away from sunlight.

The easiest way to bottle is in a pressure cooker. Otherwise, use either the slow water bath method or moderate oven method.

In a pressure cooker

1 Bring 750 ml/1½ pint water to the boil in the pressure cooker.
2 Lower the filled jars into the boiling water on to the trivet, making sure that they do not touch each other or the sides of the pan.
3 Cover and bring to low pressure (L/2.3 kg/5 lb) on a medium heat in about 10 minutes.
4 Maintain pressure for the time necessary for each fruit (see chart on page 43), then reduce the pressure slowly. Be careful when moving the cooker that the jars do not touch each other. Allow to cool as above.

Slow water bath

1 Use a deep container such as a fish kettle or preserving pan, with a false bottom (a pad of newspaper will do), so that the jars are not in direct contact with the base of the pan.
2 Fill the jars with fruit and cold syrup or water and put on the rubber bands, lids and clips or screw on the screw top lids, loosening them by a quarter turn.
3 Place the jars in the pan so that they do not touch each other or the sides of the pan, and cover with cold water and a tight fitting lid.
4 Heat slowly on the hob to the required temperature in 90 minutes, and maintain that temperature according to the fruit or vegetables (see chart on page 43).
5 Remove the jars from the water, tighten the screw bands and leave to go cold. Test the seals as above.

Moderate oven

1 Preheat the oven to 150°C/300°F/gas 2 and warm the sterilised jars before packing them with the fruit or vegetables.
2 Pour over the boiling syrup or water and cover without the clips or screw bands; jars with screw tops should be tightened and unscrewed a quarter turn.

3 Place the jars in the oven on a solid block of wood or a pad of newspaper, so that they do not touch the hot metal or each other, and cook for 30–90 minutes depending on the fruit or vegetables and size and quantity of jars (see chart below).

4 Allow a longer time for processing large numbers of jars. As soon as the time is up, tighten the lids or put on the screw bands if using. Leave to go cold and test the seals as above.

Bottling times and temperatures for 450 g–1.8 kg/1–4 lb jars

Times are approximate. For larger jars, add 15 minutes to oven bottling times or 5 minutes to water bath times.

	Minutes a moderate oven*	Minutes in a slow water bath at (temp)	Minutes in a pressure cooker
Apricots, whole	45	15 (82°C/180°F)	1
halved	55	15 (82°C/180°F)	3
Artichoke hearts, in brine	45	15 (82°C/180°F)	1
Asparagus, in brine	35	10 (74°C/165°F)	1
Blackberries	35	10 (74°C/165°F)	1
Blueberries	35	10 (74°C/165°F)	1
Cherries, whole, stoned	45	15 (82°C/180°F)	1
Currants, red/black/white	35	10 (74°C/165°F)	1
Damsons, whole	45	15 (82°C/180°F)	1
Gooseberries	35	10 (74°C/165°F)	1
Peaches/nectarines, halved	55	15 (82°C/180°F)	3–4
Pears, halved, poached	65	30 (88°C/190°F)	5
Plums/greengages, whole	45	15 (82°C/180°F)	1
halved	55	15 (82°C/180°F)	3–4
Raspberries/loganberries	35	10 (74°C/165°F)	1
Tomatoes, whole, in brine	65	30 (88°C/190°F	5

* 190°C/375°F/gas 5

Fruit with a punch

A very simple and delicious way to preserve fruit is in alcohol.
Keep the fruit whole if possible, but remove the skins and core fruit
such as pears. You can remove the stones and skins from plums,
peaches and apricots.

Pack the fruit into large jars adding 100–150 g/4–6 oz sugar for
every 1 kg/2¼ lb fruit. Fill to the top with rum, vodka or brandy.
Seal and leave to macerate for 3 months before consuming.

With soft fruits, such as blackberries and raspberries, after the 3
months is up you can either leave the slightly mushy fruits or strain
the liqueur through muslin into bottles and use the alcoholic fruit in
ice creams, sorbets, mousses or fools and enjoy the fruit liqueur as
an after-dinner digestif.

7 Jams, Jellies, Curds & Butters

To grow enough fruit to make jars of jams, jellies, curds and butters for use throughout the year is a real treat, and they are easy to make and so much better than the bought variety. Most fruits make good jams, but those with a lot of pips are often better as a jelly.

Fruit curds contain butter and eggs so do not last as long as jams and jellies (which can store for a good few years), but if they are kept in the fridge before opening they should keep for about three months, and they can be frozen. After opening, they should be eaten within a week – which won't be difficult! They are perfect in tarts and as a filling for sponge cakes.

Fruit butters have nothing to do with butter but are a preserved fruit purée that can be made from the pulp left over after making a jelly in the same way as jam. The end result is a firm preserve that can be sliced and spread. Quinces make a good butter or the dessert Spanish Membrillo (page 176).

Fruit for making preserves should be fresh, sound and not mushy, and preferably a little under ripe for optimum pectin levels. (Pectin is an enzyme found in fruit cells that helps jams to set.) Sugar is the vital preserving ingredient in a fruit preserve, but too much will make it crystallise and too little will prevent it from setting. Preserving sugar dissolves slowly causing less scum, but granulated is just as good and less expensive, and the scum can be dispersed with a knob of butter.

Pectin

All preserves (except curds) are made by boiling a quantity of fruit with an equal weight of sugar until the mixture reaches setting point. This succeeds if the fruit has the required amount of pectin and acid, which extracts the pectin, improves the colour and prevents crystallisation. If the fruit is low in pectin (such as strawberries), high-pectin fruit, such as lemon juice, is added to the mixture.

Fruits high in pectin make good jellies, as more water is added to the mix than with jam.

Fruit with high pectin

- o Apples
- o Blackcurrants
- o Blueberries
- o Citrus fruits
- o Crabapples
- o Damsons
- o Gooseberries
- o Plums (firm)
- o Quinces
- o Redcurrants

Fruit with medium pectin

- o Apricots
- o Blackberries
- o Loganberries
- o Plums (soft)
- o Raspberries

Fruit with low pectin

- o Cherries
- o Grapes
- o Peaches
- o Pears
- o Rhubarb
- o Strawberries

Equipment

Preserving pan
A large stainless steel preserving pan is ideal but any large pan will do, as long as it is big enough to allow the jam to boil rapidly without boiling over. Chipped enamel should never be used and copper can spoil the colour of red fruit. Use a long-handled wooden spoon for stirring the fruit during cooking.

Funnel
A wide-necked funnel is useful but not necessary for pouring the jam into the jars without mess.

Jars
Any jars can be used for potting preserves as long as they are completely clean. Sterilise them by leaving the clean jars in the oven on a low setting for 30 minutes. Fill the jars while they are still warm. Cover with a waxed paper disc before screwing on the lid.

Jelly bag
For making jellies and butters, you need a bowl and a jelly bag, or make your own with a large piece of fine-woven muslin or a clean (boil it first) tea towel tied between the legs of an inverted chair.

How to make jam

1. Prepare the fruit by topping, tailing, peeling, coring and weighing it. Wipe it if it is dusty, but there is no need to wash it. Stone fruits can be halved and the stones will float to the top during the boiling. Soft juicy fruits, such as raspberries and strawberries, do not need to be cooked first.

2. Weigh out the sugar – generally 1 kg/2¼ lb for every 1 kg/ 2¼ lb fruit – and place it in the oven to warm alongside the jars, or for soft fruit pour it over the fruit in a large pan and leave it to macerate for a few hours, stirring occasionally.

3. Put hard fruits in a pan, with a small amount of water, cover and cook slowly, stirring occasionally, until soft and the juice is running. With a metal spoon, skim off any stones that have floated to the top. Stir in the warmed sugar over a medium heat until it has dissolved and bring to the boil.

4   If you're making jam with soft fruits, once the juice is running, slowly bring the fruit and sugar mixture to the boil to dissolve the sugar.

5   Turn up the heat and boil rapidly for 5–15 minutes, depending on the fruit, until it reaches setting point. Test for setting point by dropping a bit of the jam on to a cold plate and letting it cool. If the jam forms a skin and wrinkles when pushed with your finger, it is ready.

6   Take the jam off the heat, skim off any scum or stir in a knob of butter to disperse it. Let it rest for 15 minutes.

7   Fill the warm jars up to the top and cover the hot jam with a waxed paper disc before screwing on the lid. Label when cold and store in a cool, dark place.

How to make jelly

1   Prepare the fruit by wiping it clean, but do not peel, core or remove any stones as they improve the setting quality and the juice will be strained off later.

2   Put the fruit and water (if needed) in a pan, cover and cook slowly, stirring and crushing the fruit to get as much juice out as possible for about 10–15 minutes.

3   Pour into a jelly bag, or similar, over a bowl and leave to drip for several hours, overnight preferably. Do not be tempted to try to squeeze more juice out of the bag as this will make the jelly cloudy.

4   Measure the juice and use 450 g/1 lb warmed sugar for each 600 ml/1 pint juice. Return to the pan with a knob of butter to prevent scum forming and to give brightness to the jelly.

5   Slowly cook to dissolve the sugar, then turn up the heat and boil rapidly until setting point is reached (see above), around 8 minutes, depending on the fruit.

6   Pot in warm, sterilised jars as for jam.

How to make fruit butter

1   If you have made jelly and have plenty of fruit pulp left, rub it through a sieve and make a butter from the purée. Or prepare fresh fruit by peeling, coring, hulling etc.

2   Stew the fruit with a very little water until soft and rub through a sieve.

3   Measure the purée and stir in 350 g/12 oz granulated sugar to each 600 ml/1 pint fruit.

4   Dissolve the sugar slowly then turn up the heat and boil rapidly until setting point is reached (see page 48). Stir occasionally to prevent it sticking and burning.

5   Allow to cool but not to go completely cold. As this is a more solid preserve, transfer it to small, rigid plastic containers. Seal with an airtight lid, label and keep in a cool, dark place.

How to make fruit curd

1   Make a fruit purée as above and measure it. Place it in the top of a double saucepan, or in a bowl over simmering water, with the required amounts of butter and warmed sugar and heat until the sugar has dissolved. For 225 g/8 oz fruit, use 225 g/8 oz sugar with 50 g/2 oz unsalted butter.

2   Add 2 beaten eggs and stir until the mixture has thickened.

3   Allow to cool a little, spoon into clean, warmed jars, cover with a waxed paper disc and seal. When completely cold, store in the fridge.

Sugar-free blackberry jam

2.75 kg/6 lb blackberries

450 g/1 lb eating apples

600 ml/1 pint apple juice

300 ml/½ pint water

Grated rind and juice of 2 lemons

Liquid pectin

Wash the fruit well. Peel and core the apples. Place in large pan and add the apple juice, water, lemon rind and juice. Bring to the boil, then simmer for about 45 minutes, stirring occasionally and scraping off the scum from time to time. Also stir once in a while to prevent it from sticking. Add the liquid pectin, according to the manufacturer's instructions, then boil for 2 minutes. Remove from the heat. Pour into clean jam jars and leave to cool and set. Keep in the fridge once opened.

8 Pickles, Chutneys & Sauces

You will never want to buy another pickle, chutney or sauce once you have grown, made and tasted your own. Pickling is a method of preserving raw or lightly cooked vegetables and fruit in vinegar (page 52), sugar, salt and spices.

Chutney is a jam-like preserve that involves cooking fruit and/ or vegetables together for a long time with vinegar and sugar acting as the preservatives. Dried fruit and spices add a special flavour. Sauces are more or less chutneys that have been sieved or puréed until smooth, then bottled or frozen.

These preserves are wonderful to have in your store cupboard as they make a delicious accompaniment to cheese, cold meats and salads. They are very versatile and you can use pretty much any of your produce to make them. Stored properly, in a cool, dark place, they can last for several years.

Pickling goes back to around 2000bc when the Mesopotamians first pickled cucumbers, and the idea of chutney – a Hindi word – was brought to Britain from India in the 18th and 19th centuries. Piccalilli, one of our most popular pickles, has turmeric and English mustard powder added to it making it that bright yellow colour we are so familiar with. Although traditionally made with cauliflower, beans and onions, piccalilli can consist of any vegetable and can represent the whole of your vegetable garden in one jar.

Equipment

As large amounts of vinegar are used in pickling and making chutney, it is important that bowls, pans, utensils and jar and bottle lids are not corrosive when in contact with the acetic acid.

Jars

Glass, wide-necked Kilner jars are best for pickles but screw top jars with plastic-lined lids can be used, as long as the necks are wide enough. Paper lids cannot be used as vinegar evaporates through them. Clean and sterilise jars by putting them in a warm oven for 30 minutes before filling.

Stainless steel pan

Chutney has to be cooked slowly for a long time so use only a large stainless steel pan, as anything else will react with the vinegar and taint the finished product.

How to pickle

1 Use only the best quality produce, which should be very fresh and washed and dried well.
2 Cut up large vegetables such as cauliflower and cucumber and leave smaller ones whole. Onions and beetroot are best pickled whole when they are young and small. Brine (page 53) or lightly cook vegetables where necessary.
3 Fruits are not brined but lightly cooked and soaked in a hot vinegar and sugar solution (about 900 g/2 lb sugar to 600 ml/ 1 pint vinegar). Soft fruits tend to go mushy so they are not worth pickling. Stone fruits that are to be pickled whole, such as damsons and cherries, need to be pricked before cooking so that they do not shrivel up. Drain and pack into sterilised jars. Continue boiling the vinegar to thicken and pour over the fruit in the jars to cover.
4 Tightly pack the sterilised jars with the prepared vegetables or fruit, being careful not to bruise or damage them. Drain off any excess water.
5 Pour in the spiced vinegar to the top. In general, it is hot vinegar for fruit, cold for vegetables, although hot vinegar can also be used for the latter. Seal jars well.
6 Leave to mature for 2–3 months before eating.

Brining

To keep their colour, shape and crispness, vegetables are sometimes first brined in a salt water solution for 24 hours. This extracts the water that would dilute the vinegar and reduce storage times. Sometimes you can skip the brining process, but the pickles will only last 3–4 months. Some vegetables, such as beans and leeks, do not need brining as they are best blanched first. You can also layer vegetables with a high water content, such as cucumber, in salt overnight, then rinse well before packing into jars. To brine with a salt solution:

1 Dissolve 225 g/8 oz coarse salt or cooking salt – not table salt, as it will make the pickle cloudy – in 2.25 litres/4 pints boiling water. Strain and cool.
2 Place the prepared vegetables in a large bowl and cover completely with the brine. Keep the vegetables submerged by weighting a plate down on top of them.
3 Leave for 24 hours. Drain and rinse well.

Vinegar

There are many vinegars you can use, but they must have an acetic acid content of at least 5%. Malt vinegar or ready-spiced pickling vinegar are fine, or you can also buy pickling spices to make your own spiced concoction. Use whole spices and herbs tied up in muslin to keep the vinegar clear and leave them to steep for several weeks before harvest time.

o **Spiced vinegar:** For a quick spiced vinegar, boil the spices tied up in muslin with the vinegar in a covered pan; remove from the heat and allow to steep for 2–3 hours. A mixture of any of the following can be used: cinnamon stick, peppercorns, coriander seeds, bay leaf, cloves, blades of mace, allspice berries, dill seeds, dried chillies, mustard seeds.
o **Herb vinegar:** Half-fill a bottle with fresh sprigs of either thyme, sage, rosemary, tarragon or dill, or with a selection, and pour on a colourless distilled malt vinegar or white wine vinegar and leave to steep for 2–3 weeks.
o **Sweet vinegar:** Add 450 g/1 lb sugar to 2.6 litres/4½ pints malt vinegar and bring slowly to the boil with spices such as cinnamon sticks, allspice berries, star anise and fresh ginger tied up in a muslin bag until the sugar has dissolved. Turn off the heat and leave to infuse overnight.

How to make chutney

You can make chutney from all sorts of odds and ends of fruit and vegetables and they don't have to be in tip-top condition. For different flavours and colours, you can mix vinegars, such as half and half red wine vinegar and balsamic, and sugars, such as half and half muscovado and white granulated, or try golden syrup instead. For a rough guide, for every 2 kg/4½ lb of fruit and vegetables, use 600 ml/1 pint vinegar and 450 g/1 lb sugar.

1 Prepare the vegetables and/or fruit and chop into small pieces. Some watery vegetables, such as marrow, will need to stand for 24 hours sprinkled with 30 ml/2 tbsp salt and then be drained.

2 Place in a large pan, preferably stainless steel, with the vinegar of your choice, sugar, dried fruits, if using, and spices.

3 Bring to the boil, turn down the heat and simmer gently for about 2 hours, stirring occasionally, until the chutney has thickened and reduced to a jam-like consistency and all the liquid has evaporated.

4 Take off the heat and allow to cool a little. It will continue to thicken as it cools.

5 Spoon the hot chutney into warm, dry, clean jars and seal and label them.

6 Store in a cool, dry, dark place for a few months for the flavours to blend and mature before eating. Chutney will keep in a coo, dry, dark place for at least 2 years.

How to make sauces

Sauces contain the same ingredients as chutney to make them last. If you are making a large batch, you can keep half for chutney and the other half to make into a sauce to serve with a variety of meat dishes and the full English breakfast.

1 Make chutney as above and allow to cool completely.

2 Sieve the mixture into another pan, using a non-metallic sieve.

3 Reheat and pour into sterilised sauce bottles with screw-top vinegar-proof lids.

4 Store in a cool, dark, dry place for a few months before using.

9 Juices & Cordials

Turn your vegetables and fruit into refreshing, healthy drinks and store them in the freezer until you need them. Cordials can be kept in sterilised bottles.

If you make juice soon after picking, you will have a drink full of vitamins and nutrients. Juices are great health-builders and if you drink them regularly, you should not only feel healthier, but also acquire more energy and build up resistance to disease.

Fruit cordials are made with a sugar syrup and are diluted with water – still or sparkling – to make a cool and refreshing drink, or hot as a soothing nightcap. Cordials can also be served poured over ice cream or added to a fruit salad.

Equipment

Juicer

To make juice you will have to make an initial investment of an electric wholefruit juicer. You will then need to spend no more money and juice-making will become quick and easy to do. Many wholefruit juicers have necks wide enough to take whole apples and pears, so that all you have to do is wash the fruit before feeding it into the machine. They also remove every last bit of juice just leaving the dry pulp. However, if you have a large number of apple trees it may be worth investing in an apple press. If you don't want to buy one, you can make one with two pieces of board attached by a G-clamp at each corner, then place a wider container underneath to catch the juice that spills over the sides. Alternatively, many fruit farms make their own juices these days, so maybe they will press your fruit for you.

Equipment for cordial

To make cordial you will need:

- o Earthenware casserole
- o Saucepan or preserving pan
- o Jelly bag or large piece of muslin
- o Bottles, sterilised in a low oven for 30 minutes

Containers

Juice can be stored in the freezer for several months, packaged in foil freezer bags or used plastic bottles. For cordial, use clean glass.

How to make your own juice

Any fruits can be blended together into juices, a wide variety of vegetables too, but if you are making juice for health purposes, it is said that fruits and vegetables should not be combined as their enzymes are not compatible, except with apples.

To prepare

- o Wash all fruits and vegetables thoroughly, using a vegetable scrubbing brush if necessary. Slice or dice them immediately before they are to be juiced and drunk or frozen to minimise vitamin C loss. Dip those that discolour quickly into acidulated water (5 ml/1 tsp lemon juice to 300 ml/½ pint water, or 50 g/2 oz salt to 4.5 litres/1 gallon water), drain and pat dry with kitchen paper.
- o Most fruits and vegetables can be juiced with their peel, seeds and stems to obtain optimum nutrition; stone fruit, such as peaches and plums, should have their stones removed first.
- o Use a salad spinner to remove excess water from green vegetables.

To juice

Follow the manufacturer's instructions for your juicer. After juicing, pass the liquid through a fine sieve if it still has bits in it, then pour it into foil freezer bags or recycled, well-washed plastic bottles – leave space at the top for ice expansion – and freeze immediately, if not consuming. Defrost the juice in the fridge overnight and drink within 2 days. You can keep the juice for longer if you sterilise it using one of the bottling methods described on pages 41–43.

Juicing tips

o Do not force the produce through the juicer: feed it in slowly.
o Bunch greens together in clumps and put them through with the pusher.
o Cut root vegetables into chunks.
o With some produce, such as apples and carrots, the remaining pulp can be used in other recipes.

Vegetables suitable for juicing are:

o Beetroot
o Cabbages and other greens
o Carrot
o Celery
o Fennel
o Parsnip

Strong flavoured vegetables such as beetroot and greens are better combined with apple or carrot.

Most fruits can be juiced but the best are:

o Apple
o Blackberry
o Blackcurrant
o Blueberry
o Grape
o Loganberry/boysenberry
o Peach
o Pear
o Raspberry
o Redcurrant

Again, apple and carrot are the best diluters.

How to make cordial

Any ripe soft fruit can be used to make cordial and many of them combine well. Other flavourings can also be added, such as ginger and cinnamon.

The best combinations are:

o Blackberry and apple
o Elderberry and gooseberry
o Peach and raspberry
o Raspberry and redcurrant
o Strawberry and rhubarb

1 Put any quantity of fruit into an earthenware casserole with 150 ml/¼ pint water.
2 Cover and place in the oven preheated to 150°C/300°F/gas 2 for 1 hour until the juice is running.
3 Tip into a jelly bag over a large bowl, or into muslin tied between the legs of an inverted chair, to strain the juice overnight. If you squeeze the bag the juice will go cloudy.
4 Measure the liquid and add 350 g/12 oz sugar to each 600 ml/1 pint juice.
5 Boil the sugar and juice together rapidly for 5 minutes, skimming off any scum. Allow to cool.
6 Pour into warm, dry, sterilised bottles and store in a cool, dark, dry place, or freeze in small quantities in polythene containers. Once opened, keep in the fridge for up to 3 months.

10 Cider & Wine

Wine can be made from a wide variety of produce. In fact, most of what is growing in your garden – as long as it is not poisonous – can be turned into a palatable wine with a little care and attention and some patience. Try beetroot, gooseberries, parsnip, blackberries, even broad beans… the list is endless. If you have a surplus of apples – and it only takes one mature tree to provide a few sackfuls – you can brew up some cider, which is more straightforward than making wine and you can drink the results much sooner.

Pears can also be made into cider – called perry – but they have to be the hard perry pears to obtain a good flavour. However, you do not have to grow cider apples unless you want a traditional cider or scrumpy. Cooking apples mixed with dessert varieties can make a delicious light cider similar to a dry white wine.

Equipment

Once you have made the initial outlay, there is little more equipment that you will need to buy. All equipment in wine and cider making must be scrupulously clean and sterilised in a solution of 5 ml/1 tsp of sodium metabisulphite, or 2 Campden tablets, in 600 ml/1 pint hot water, with a few crystals of citric acid to help release the active agent, sulphur dioxide.

Juice extractor

For both wine and cider-making you need to extract as much juice as possible from your fruit and vegetables. You may prefer to invest in an electric wholefruit juicer, which will do the job quickly and efficiently, or a fruit press, which will be more labour intensive but more suitable for large quantities of apples (page 57).

Fermentation bucket

For the initial fermentation process in both wine and cider making, you will need a high-density polythene or polypropylene bucket with a lid, for soaking and mashing soft fruit, flowers or berries. It is safer to buy one specifically for brewing to be sure that there are no unpalatable chemicals in the plastic that can leach into the liquid. For cider, you only need a food-grade plastic keg or barrel.

Fermentation jars

A couple of 4.5 litre/1 gallon glass fermentation jars, or demijohns, with a narrow neck that can hold a cork or rubber bung are important for the fermentation stage.

Fermentation lock

The fermentation lock is a simple device inserted into the cork or rubber bung stopping up the fermentation jar that allows gas to escape from the fermenting cider or wine, while keeping out air, wild yeasts and fruit flies.

Plastic tubing

A length of plastic tubing is needed to siphon the fermented liquid out of the jar without disturbing the sediment at the bottom.

Hydrometer

A hydrometer is worth investing in if you want to be sure that your wine or cider will be successful. It is an instrument that measures the sugar level in the juice so that you know how much sugar to add for correct fermentation to take place.

Acid indicator papers

For vigorous fermentation, the liquid base for the wine, called a must, should be slightly acid. Many fruits contain enough acid naturally, but you will need to add some to a number of recipes in the form of orange juice, lemon juice or powdered citric acid.

Alternatively, too much acid can be countered by the addition of precipitated chalk. To find out the acid level, you can test the must by dipping an acid indicator paper into it, sold in small books or rolls, to give you the pH level.

Bottles

Use proper wine bottles for wine as they are strong enough to withstand any pressure that might be caused by unexpected fermentation later on. Used ones are fine, but they must be properly cleaned and sterilised first using sodium metabisulphite (page 59). Dry completely before filling.

It is very important that sparkling wine is stored in tougher champagne-style bottles with metal clips as fermentation continues in the bottles and can cause explosions. Red, rosé and other dark-coloured wine should be put into dark bottles to prevent the colour of the wine fading. Cider can be bottled in any type of bottle with a screw top.

Clearing agents

If your wine is slow to clear, it might be necessary to add a fining agent, pectinase or amylase, or to strain it through a filter (page 65).

Corks

New corks should be used so that they do not contaminate the wine. Soak them in cool water for 24 hours first, or in hot water for half an hour. Weight them down so that they stay submerged. After they have been soaked, sterilise them in a sodium metabisulphite solution as above. Fit the corks to the bottle with a corking tool for an airtight fit.

For sparkling wine, use hollow plastic stoppers and fit a wire cage (muselet) over the top.

How to make wine

You can get as much pleasure out of making wine as drinking the result, starting with growing the produce – whether it be grape, rhubarb, marrow, parsnip or damson – harvesting it, then watching it slowly ferment and become alcoholic in a bubbling mixture of sugar and yeast.

Yeast starter bottle

Your must will begin to ferment more quickly and reliably if you start the yeast fermenting in a sterilised starter bottle first. You will need:

300 ml/½ pint fruit juice

25 g/1 oz sugar

Pinch of citric acid crystals

5 ml/1 tsp winemaker's vitaminised yeast nutrient

1 sachet of granulated dried wine yeast or a yeast tablet

1   Boil all the ingredients together, except the yeast, until the crystals have dissolved.

2   Take off the heat and cool until lukewarm – about 21°C/70°F.

3   Add the yeast according to the packet instructions.

4   Pour the solution into the starter bottle, plug with cotton wool and leave to stand in a warm place.

5   After a short while, the liquid will begin to bubble; it is ready to be added to the must.

1   Prepare the must by extracting the juice from the produce you have chosen with a juice extractor or fruit press. For hard root vegetables and fruit, gently boiling until the produce is soft is an effective way of extracting colour and flavour. Another way is to steep the chopped up produce in a bucket of cold water for several days with 1–2 Campden tablets crushed and stirred in to destroy any wild yeasts

or harmful bacteria. Alternatively, add boiling water to the bucket (which negates the need for Campden tablets). A quantity of chopped raisins may be added now to give body, or vinosity, to the wine. Cover the bucket well.

2 Produce that contains pectin needs the addition of 5 ml/1 tsp of the pectin-destroying enzyme pectinase (page 65).

3 Transfer the must into a clean fermentation bucket, straining it through a fine mesh sieve, if necessary. Water can be added at this stage, if your recipe requires it.

4 Test the acidity level of the must with an acid indicator paper. You should aim for a pH of 3 or 4. Add lemon juice or citric acid if necessary or precipitated chalk in small amounts if it contains too much.

5 Take a hydrometer reading and add the necessary amount of sugar to arrive at a reading of approximately 1.100 for a dry wine or 1.120 for a sweet one (or follow your recipe if you don't have a hydrometer). The optimum quantity of sugar is approximately 1.1 kg/2½ lb white granulated sugar for every 4.5 litres/1 gallon liquid (250 g per litre/8 oz per 25 fl oz) added in stages (half at the beginning, the rest in smaller quantities as fermentation proceeds).

6 Make sure the temperature of the must is between 19–24°C/66–75°F before adding the yeast starter (page 62). Re-cover the bucket securely.

7 Fermentation will begin quite violently at first. Leave for 4–5 days.

8 Strain the fermenting liquid into a sterilised fermentation jar or demijohn, leaving enough room for any more vigorous bubbling to take place without overflowing. Seal with the fermentation lock half filled with water containing ⅛ Campden tablet to prevent harmful bacteria from ruining fermentation. Label and leave in a warm place at a constant temperature 21–25°C/70–77°F for several months. When the carbon dioxide gas has stopped escaping in bubbles through the lock, it is ready for racking.

9 If fermentation stops after a short period, check that the temperature is correct and constant. If necessary, add another 5 ml/1 tsp of vitaminised yeast nutrient or citric acid to get it back on its way.

Racking and clearing

During the first few weeks or months, the must will be a thick, opaque liquid, with a suspension of particles of yeast and pulp. As the yeast is used up in producing alcohol, the debris (lees) falls to the bottom and the clearer wine that is left has to be siphoned off (racked) through clear plastic tubing.

1 Stand the fermentation jar to be racked on a table and place another sterilised jar on the floor beneath it.
2 Remove the fermentation lock and bung and put one end of the tubing halfway into the liquid. Gently suck the other end of the tubing.
3 As the wine fills the tubing, tightly pinch the end you are sucking and place it inside the empty fermentation jar. Allow the wine to flow into the jar. Make sure that the wine level of the original jar does not fall beneath the tubing, but remove as much as you can without disturbing the lees.
4 When all the clear wine is inside the new jar, add a crushed Campden tablet, top with a solution of sugar and water in the same proportion as you did originally (see Step 5 on page 63), leaving space at the top, and fit a clean fermentation lock, as before.
5 Repeat at monthly intervals until the wine is clear and has finished fermenting. Move to a cool place.

Bottling and storing

When the wine has been racked several times and is crystal clear, it is ready to be bottled.

1 Fill the clean, sterilised bottles with wine, leaving a small space at the top.
2 Seal with a sterilised cork.
3 Label with the name and date.
4 Store the bottles on their sides, so that the corks remain in contact with the wine and stay moist.
5 Leave to mature, undisturbed, for several months in a dark, cool place at an even temperature of about 13°C/55°F.

When the wine won't clear

o If your wine remains cloudy, you will have to try to clear it artificially with finings, a substance containing isinglass, gelatine, casein and egg white. Another fining agent is bentonite, a powdered clay mineral. Follow the manufacturer's instructions to achieve the best results.

o Cloudiness can be caused by too much pectin, which can happen if the wine has been made by boiling fruit that contains a lot of pectin (page 46). It can be treated by adding pectinase. However, it is best to prevent this happening in the first place by adding the pectinase at the early stages when the mashed pulp has cooled and before the sugar is added. To find out whether pectin is an offender, add a small amount of the wine to a small bottle of methylated spirits and shake vigorously. Leave the mixture to settle for some time and if jelly-like strings or blobs appear, pectin is present.

o Cloudiness can also be caused by too much starch in fruit and vegetables, such as unripe apples and parsnips, if the produce is boiled too rapidly to begin with. This can be cleared with a starch-destroying enzyme called amylase.

o If all else fails, try filtering your cloudy wine through filter paper, folded to fit into a funnel, or with a filter kit.

How to make cider

There are several ways to make cider with apples or pears. The traditional method is by allowing the apples to ferment with the wild yeasts from the air and natural sugars from the fruit. This results in a cloudy alcoholic drink that can be quite tart.

For a more refined cider, use wine-making champagne yeast and extra sugar and a method that is very similar to making wine (Sparkling Apple Wine, pages 72–74). For a sweet cider, use only dessert apples, but if you want to tone it down a little, a mixture of cooking apples or crabapples (about 10–20%) with the dessert ones is best. The fruit must be ripe.

Pear cider, or perry, is best made with perry pears.

About 4.5–6 kg/10–14 lb apples or pears will make 4.5 litres/1 gallon cider.

Traditional cider

1 Chop and macerate the fruit to make a pulp and extract as much juice as possible, using a juicer and press. Wrap the pulp in muslin or similar when pressing it.

2 Fill a clean, sterilised keg or barrel right to the top with the juice. Close the top but leave the bung out so that the natural yeasts can enter. Leave to ferment for several weeks. You should be able to see it fermenting after a day or two when the froth starts coming up out of the bung hole.

3 When fermentation stops after several weeks, replace the bung and leave to mature for at least 8 months.

PART 2

A to Z
of storing fruit
& vegetables

Storing

Drying ☀

Freezing ❄

Bottling ▦

Preserving

Juicing ◌

Wine making ⧗

A APPLES 🖼 ☀ ❄ 🌢 🗒 🍸

There are at least 5,000 named varieties of apple and a great many different ways of storing them so that they can be enjoyed throughout the year. Just one mature tree in your garden can provide a large quantity of apples that can keep you going throughout the year and if you are lucky enough to have an orchard, processing the fruit into delicious concoctions to store and use is very rewarding.

Apples are a versatile fruit and divide into two types, cooking and dessert. Cooking apples are tart and acidic and break down during cooking, while dessert apples are more crisp and sweet. All apples freeze well, either open frozen (to be cooked later) or cooked – simply stewed or as a purée. Most will provide a good juice and you don't need cider apples to produce a quality cider.

Varieties

If you are planning to grow apple trees, bear in mind that the late fruiting dessert varieties that crop in October store best and can last until spring. Many varieties can make a good cider, especially a mix of cookers (Bramley's Seedling) and dessert apples (Tom Putt and Blenheim Orange), and most varieties of sweet apple, such as Jonagold, Discovery and Worcester Pearmain, make delicious juice. The most popular cooker, Bramley's Seedling, sweetens during storage. Crabapples are high in pectin and make good jelly.

Apples that store well
- **Ashmead's Kernel:** Late cropper old dessert russet with sweet peardrop flavour; makes good cider; store until February.
- **Blenheim Orange:** Dessert apple with sharp, nutty flavour with large fruits that cook well; harvest early October; store until January.
- **Braeburn:** Popular eating apple, sweet, crisp and juicy; excellent juice; cropping in October; store until March or April.
- **Bramley's Seedling:** Classic British cooker; harvest in October; store until the spring, growing sweeter by the day.
- **Brownlees Russet:** Excellent late apple, improves on keeping; pick in October; store until January/February.

o **Jonagold:** Popular flavoursome variety that makes good juice; harvest mid-October; store until February.

o **Red Falstaff:** Produces a heavy yield of bright red apples with an excellent flavour; can be grown in pots; pick early October; store until March.

Early cropping apples

o **Tom Putt:** Old Dorset cider apple with sharp taste; good for cooking as well as for making cider and a medium sweet apple juice; crops September.

o **Worcester Pearmain:** Makes excellent sweet juice; crops early September; does not store.

o **Discovery:** provides medium sweet juice; crops in August; does not store.

Cooking and preparing

Dessert apples keep their shape better when cooked and are perfect in apple tarts, especially those with red skins to add some colour. Cooking apples cook to a mush so are best enjoyed in sweet, comforting crumbles and pies and marry perfectly with pork, either on the side as a sauce or in casseroles. Sliced apples can be fried in butter to partner game and other rich meats.

o **To harvest:** Pick from the tree when ripe, by holding the apple in your hand and giving it a gentle twist; if it is ready it will come away easily. Be careful not to shake other apples from the tree. Store the perfect fruit and cook the windfalls and blemished ones.

o **To prepare:** Peel, core and slice apples for cooking, cutting out any bruises and other blemishes. Immediately set aside in a bowl of acidulated water (5 ml/1 tsp lemon juice to 300 ml/½ pt water) until needed, to prevent the flesh from turning brown.

o **To stew:** Prepare the apples and cook over a gentle heat, covered, in a small amount of water with sugar to taste until pulpy. Cookers will soften quicker than dessert apples.

A Storing 🔲

Dry-storing on trays or wrapped up in newspaper in boxes is the simplest method of keeping apples, as there is no peeling and coring involved. However, you do need some space, depending on how many apples you have to store. It is important to pick perfect specimens off the tree before they fall off and bruise, and to handle the fruit very carefully.

Drying ☼

Apples dry well to make a tasty nutritious snack and rehydrate by soaking in cold water overnight, to be added to winter fruit salads and cakes, bakes and casseroles. Peel and core the apples leaving them whole. Slice them thinly into rings, dipping or brushing them in lemon juice or acidulated water (page 69) to prevent them from going brown. Pat dry and air dry.

Freezing ❄

You can freeze apples cooked or raw, but you have to cook the latter after thawing as the texture becomes soft and sponge-like. The advantage of freezing apples is that you can use windfalls, just cut out the bad bits and bruises when you are peeling them.

o **Open freeze:** Peel, core and cut into quarters, then rub all over with the cut side of a lemon. Lay them on an oiled baking tray and freeze. When frozen, tip into freezer bags and return to the freezer.

o **Purée:** Stew the apples as above. Cool and process in a blender or pass through a sieve. Pour into freezer bags or rigid containers and freeze.

o **Stewed:** Prepare and cook as for purée but don't blend or sieve the cooked apples. Freeze in bags or rigid containers when completely cold.

o **To thaw and use:** Heat straight from frozen or thaw overnight in the fridge. Use frozen raw slices of apples in pies and crumbles as they will create more juice as they cook.

o **Storage time:** 8–12 months.

Juicing 🔘

The sweeter the apple the more delicious the juice. If you have a large number of apple trees and you have wheelbarrows full to make into juice, it may be worth investing in an apple press. Or a nearby a fruit farm may press them for you. Otherwise use a juicer and freeze the result.

Preserving ▨

Cooking apples are high in pectin and are traditionally paired with other low-pectin fruit in jams and jellies, the favourite being Blackberry and Apple Jam (pages 94–95). Apple Jelly makes a good base for mint or ginger.

Apples can be teamed with plums to make a delicious butter.

—— Apple and Plum Butter ——

Makes about 2 kg/4½ lb

1.5 kg/3 lb apples
450 g/1 lb plums, stoned
Granulated sugar

1 Peel and core the apples and cut into slices.
2 Cook in a very little water until soft.
3 Add the plums and cook until soft.
4 Rub the mixture through a sieve and measure the purée.
5 Stir in 350 g/12 oz sugar to each 600 ml/1 pint purée.
6 Bring to the boil and boil hard to setting point, stirring occasionally to prevent it from sticking and burning on the bottom.
7 Allow to cool a little and spoon into rigid plastic containers.

Apples can be the sole fruit in a traditional apple chutney, teamed up with onions and spices. They also pair well with tomatoes and other fruit and will keep in the store cupboard for several years.

Traditional Apple Chutney

Makes about 2 kg/4½ lb

1 kg/2¼ lb cooking apples, peeled and chopped

225 g/8 oz onions, peeled and finely chopped

225 g/8 oz sultanas

25 g/1 oz mustard seeds

50 g/2 oz salt

2.5 ml/½ tsp freshly ground pepper

225 g/8 oz granulated sugar

750 ml/1¼ pints vinegar

15 ml/1 tbsp ground ginger

1 Put all the ingredients into a pan and bring to the boil.

2 Simmer until thick and brown, stirring frequently.

3 Allow to cool a little and spoon into warmed, sterilised jars. Seal and label.

4 Store in a cool, dark place for a few months before using.

Cider and wine making 🍷

Making your own cider the traditional way is very straightforward. You don't even have to wash the apples first, as the natural yeasts on the apple skins are an important ingredient. You can also make a more refined cider or sparkling wine with your apples.

It is important to use champagne bottles, hollow plastic stoppers and wire cages (muselets) for bottling sparkling wine in order to contain the fermentation that continues in the bottle and to avoid any nasty explosions.

Sparkling Apple Wine

Makes about 2 litres/5½ pints

2 kg/4½ lb dessert apples, washed and finely chopped

225 g/8 oz raisins, chopped

1.1 kg/2½ lb granulated sugar

juice and grated zest of 1 lemon

5 ml/1 tsp pectinase

1 sachet wine yeast

5 ml/1 tsp vitaminised yeast nutrient

1 Campden tablet

A

1 Put the apples and raisins into a fermentation bucket and pour in 2.25 litres/4 pints cold water.

2 In a pan, dissolve the sugar in 1.75 litres/3 pints boiling water. Add it to the bucket and leave to cool.

3 When it is lukewarm, add the lemon juice and zest, the pectinase, the wine yeast and vitaminised yeast nutrient, followed by the Campden tablet.

4 Cover the bin securely and leave to ferment for 8–9 days, stirring and mashing the pulp daily.

5 Strain the liquid off through a fine mesh sieve or straining bag into a fermentation jar. Top up with cold water if required and fit a fermentation lock containing sterilising solution.

6 Put the jar in a warm place until fermentation is complete and the wine begins to clear, when the sugar will have been used up and a dry wine produced.

7 Rack the wine at suitable intervals until it is perfectly clear (page 64). Then leave the wine to mature.

8 After about 6 months, dissolve 65 g/2½ oz sugar in each 4.5 litres/ 1 gallon of wine.

9 Prepare a yeast starter bottle (page 62) using champagne yeast and when it is fermenting vigorously add it to the wine and seal the jar with a fermentation lock.

10 When the wine is fermenting again, siphon it into champagne bottles filling them to within 5 cm/2 in of the top, seal with hollow plastic stoppers and fit a wire cage (muselet) over each one, tightening them firmly over the mouth of the bottle.

11 Label and date. Put the bottles in a warm place for about a week so that the fermentation can continue, then move to a cool, dark place to be stored upside down for a year. This enables the yeast deposit to collect inside the hollow stopper.

12 When the sparkling wine is ready for drinking, place the neck of the bottle into some crushed ice to freeze the sediment inside the stopper. Remove the wire cage and ease out the stopper, holding the bottle horizontal.

13 As the secondary fermentation will have used up all of the sugar, the wine will be very dry, so either sweeten each bottle with a saccharin

A

tablet or make up a solution of 450 g/1 lb caster sugar dissolved in 600 ml/1 pint wine and add to the bottles to suit individual taste.

14 After sweetening, seal the bottles again with a clean stopper and rewire it until the wine is required, perhaps an hour or so later.

15 Serve the sparking apple wine at a temperature of about 8°C/45°F.

APRICOTS

If you have a south-facing wall or a large conservatory or greenhouse it is worth growing apricots as they are wonderful dried, bottled, in conserves and they freeze well. Although the trees are usually hardy, it is the blossom that suffers from the frost, so you need a good warm spring with no surprise late frosts to get a large enough crop for keeping.

Varieties

o **Flavourcot:** A new Canadian variety that blossoms late, avoiding the frosts.

o **Goldcot:** An early fruiting variety, claimed to be very hardy and less sensitive to the cold.

o **Moorpark:** Easily available and does best grown along a wall, fruiting in July or August.

o **Tomcot:** A self-fertile French variety bearing large fruit that is also claimed to be very hardy, coping with the cold.

Cooking and preparing

Apricots should be picked when they are ripe and plump with juice and eaten straight from the tree, but they are also delicious halved, stoned and baked in the oven with a sweet almond topping, cooked in tarts and in savoury stuffings, particularly for lamb and turkey. Apricot jam is a perfect ingredient for layered sponge cakes. Apricot purée makes a good base for sorbets, ice creams and mousses.

o **To prepare:** Choose blemish-free fruit, unless you are cooking them first or using them for jam, in which case, cut away any blemishes. Wash, halve and stone them. For some recipes you may need to remove the skins: plunge the fruit into a large bowl of boiling water for 1 minute, drain and quickly slip off the skins.

o **To purée:** Prepare the fruit and remove the skins, then poach gently in a small amount of water and sugar according to the sweetness of the fruit, until soft and mushy. Blend or sieve to a purée.

Drying ☼

A supply of your own dried apricots is a treat to have in the store cupboard, but it is only worth drying blemish-free fruit that are just ripe. Wash, dry them well and remove the stones. Lay them out on wire racks and choose a drying method. You could also string them up.

Freezing ❄

Apricots freeze well whole, puréed and cooked in syrup. Select ripe fruit with unblemished skins. They should give slightly when squeezed but should not be too soft.

o **To prepare:** Wash, dry, halve and de-stone the fruit. Brush the flesh with lemon juice to prevent discoloration.
o **Dry-packing:** Freeze in rigid containers with 100 g/4 oz sugar for every 450 g/1 lb fruit.
o **Purée:** Process the fruit as above and when cold, pour into freezer bags in usable quantities.
o **In syrup:** For every 450 g/1 lb fruit dissolve 100 g/4 oz granulated sugar in 300 ml/½ pint boiling water. Allow to cool and pour over the apricots in rigid containers, leaving some space at the top.
o **Poached:** Poach the prepared fruit halves in the syrup (as above) for 2–3 minutes until just tender but still holding their shape. Leave to cool and freeze with the syrup.
o **To thaw and use:** Take out what you need the night before and defrost in the fridge, or at room temperature. You can cook the purée from frozen.
o **Storage time:** 12 months.

Bottling ▣

The flavour of apricots responds well to bottling. They may be bottled whole or halved with the stones removed. Use firm, blemish-free fruit that are only just ripe; remove the skins and bottle in syrup, adding 15 ml/1 tbsp lemon juice to prevent discoloration, and pack into jars.

A Juicing 🔾

Apricots don't usually juice very well as they are generally quite dry, but they make a very good smoothie. Either use frozen purée as above and dilute with another juice or water to the consistency you prefer, or purée the fruit soon after picking, mix with another juice or water and freeze in plastic bottles. If you are using frozen purée or juice, do not refreeze.

Preserving 📧

Apricots make very special jam. They contain medium levels of pectin, so should set without much difficulty.

ARTICHOKES, GLOBE ❄️ ▫️ 📧

The globe artichoke is a close relation of the cardoon, a large ornamental thistle that is often grown in a wide flower border. It is a beautiful looking plant but needs plenty of space as it can grow up to 1.8 m/6 ft with a 1.5 m/4 ft span.

The impressive green or purple flower buds contain the edible parts, which are the base of the bracts (bud leaves) and the fleshy, flavoursome heart hidden beneath a tuft of inedible hairy fibres called the choke. Although best eaten fresh, with a vinaigrette or hollandaise sauce or just melted butter, globe artichokes can be frozen, cooked or raw, bottled or pickled. However, it is only worth preserving the hearts, so there is a lot of wastage.

Varieties

o **Green Globe Improved:** A large plant that grows up to 1.8 m/ 6 ft and produces many good quality buds for harvesting.
o **Violetta di Chioggia:** A stunning plant with very tasty purple buds that grows to 90 cm–1.2 m/3–4 ft, making it ideal for the flower border or large tub.

Cooking and preparing

A

Some people may feel that it is too much of a palaver to eat globe artichokes; but it can be quite fun as, to begin with, fingers are the only utensils you need. The fibrous stalk is edible, as well as the heart and base of the bract. Cut the buds a couple of inches down the stalk when the bracts are just starting to open out from the bottom but the top is still closed up. Artichokes will keep for several days in the salad compartment of the fridge in a plastic bag with some holes cut into it.

o **To prepare:** Rinse and remove the darker outer leaves at the base, trim the stalk and thorny bract tips, if you prefer, and with a sharp knife, slice off the top to leave a tight bulb.

o **To boil:** Cook whole in boiling, salted water, with a squeeze of lemon juice to prevent browning, for at least 20 minutes, depending on the size, until the bracts pull off with a slight tug. If they fall away too easily, they are overcooked. Drain upside down in a colander.

o **To bake:** Wrap each artichoke in foil, sprinkled with lemon juice, and bake in the oven at 200°C/400°F/gas 6 for about half an hour until the bracts can be pulled off quite easily. The fleshy base should be soft but not mushy.

o **To eat:** When they have cooled enough, one by one pull off a bract with your fingers, dip the fleshy base into a warm, vinaigrette, hollandaise sauce or similar, and scrape it off with your teeth and into your mouth. Eventually, you will be left with the inedible hairy, fibrous choke, which you must cut out carefully with a sharp, pointed knife and discard to reveal the delicious, sweet-tasting heart.

Freezing ❊

Globe artichokes are best cooked before they are frozen; if they are not too large, the buds can be frozen whole or you can open freeze just the hearts after cooking. Be careful not to overcook them.

o **To prepare:** Remove some of the coarse outer leaves and trim the sharp ends off the rest. Cut off any stalk to about 2.5 cm/1 in.

A

o **Uncooked:** If you prefer to freeze them raw, blanch them first by boiling them in water with a good squeeze of lemon juice added, to prevent browning, for 6–8 minutes, then plunging them into a bowl of iced water to cool quickly. Drain well upside down and pat dry with kitchen paper. Open freeze.

o **Cooked:** Prepare and cook the artichokes as above. Drain upside down in a colander and pat dry with kitchen paper. Place on an oiled baking tray and freeze. Once frozen, pack in boxes or freezer bags and return to the freezer. As they can cross taint, cover the boxes with an extra layer of clingfilm before snapping the lid on, or use two freezer bags to make a double thickness.

o **To thaw and use:** Cook the blanched artichokes from frozen in boiling, lightly salted water until a bract pulls out easily. Plunge the frozen cooked artichokes into boiling water and simmer until heated through. Drain upside down in a colander. Use as fresh.

o **Storage time:** 6 months raw, 3 months cooked.

Bottling 🔲

Cooked artichoke hearts are excellent bottled in brine with herbs and garlic, and can be used as antipasti or on pizzas or just straight from the jar.

o **To prepare:** Cook in boiling water with a little lemon juice squeezed into it until the bracts pull out quite easily. Remove all the bracts and cut out the choke with a sharp pointed knife and discard. You may prefer to grill the cooked hearts either on a hot barbecue or griddle pan but it is not essential. Allow them to cool and dry completely with kitchen towel.

Preserving 🖼

Pickle the hearts in a mild herb vinegar so as not to overpower the delicate taste of the artichoke hearts. Prepare and cook the artichokes, being careful not to overcook. Pack the hearts into a large, sterilised jar, and fill with vinegar.

ARTICHOKES, JERUSALEM 🖸 ❄   A

These knobbly little root vegetables, related to the sunflower, are very easy to grow and come back year after year; it only takes a tiny piece of tuber to be left in the ground to make a new plant the following year and if you are not careful they will take over your vegetable patch, so ideally give them a space of their own.

Unrelated to the globe artichoke, Jerusalem artichokes can provide you with sustenance from late autumn and throughout the winter, by just leaving them in the ground and digging them up when you need them, heavy frosts permitting, but it is best to freeze them cooked.

Varieties
o **Fuseau:** The most widely available, this variety is popular as it is the least knobbly; it is also the highest yielder.
o **Gerard:** A pink-skinned cultivar with not quite such a high yield as Fuseau, but some say it is tastier.

Cooking and preparing
Jerusalem artichokes make wonderful soups, purées, mashes and bakes that can all be frozen.

o **To prepare:** Wash the tubers well and peel as thinly as possible, cutting off any knobbly bits that are difficult to peel around. Plunge them straight into a bowl of cold water with a squeeze of lemon juice added to prevent any discoloration. Drain, slice and cook. If you prefer, the skins can be kept on, so they just need a thorough scrub.
o **To boil:** Cook in salted boiling water for about 10 minutes or until just tender; they can go mushy quite quickly.

Storing 🖸
Leave the artichokes in the ground until you want them, covering them with straw at the threat of hard frosts. Or dig up enough for a few meals before a cold snap is forecast, wash and dry them and keep them in the fridge in a plastic bag for up to 2 weeks. If you have enough tubers that are blemish free or that you haven't chopped a bit off when digging them up, you can dry store them in sand to keep for a few months.

A Freezing ❄

Jerusalem artichokes are best frozen after they have been cooked, as a purée or a mash.

o **To prepare:** Wash and peel the tubers, digging out any bad bits. Place them in a bowl of cold water with a good squeeze of lemon juice added to prevent discoloration.

o **Purée:** Slice the tubers and boil them in vegetable stock or water for 10–15 minutes or until tender. Drain, reserving some of the liquid, and mash, or purée in a blender; add some of the reserved liquid if the mixture is too thick. Freeze in freezer bags or rigid containers.

o **Mash:** Boil the prepared artichokes in water and mash before freezing in rigid containers or freezer bags.

o **To thaw and use:** Reheat from frozen, stir in some butter or cream and freshly ground black pepper or make into a soup.

o **Storage time:** 3 months.

Jerusalem Artichoke Soup

Serves 4

450 g/1 lb Jerusalem artichokes, peeled and sliced, or frozen purée, thawed

15 ml/1 tbsp olive oil

1 large onion, peeled and chopped

900 ml/1½ pints vegetable stock

25 g/1 oz butter

15 ml/1 tbsp plain flour

300 ml/½ pint milk

Freshly ground pepper

Chopped fresh parsley

1 Place the sliced artichokes in a bowl of cold water with a squeeze of lemon juice added and set aside.

2 Heat the oil in a large pan and gently fry the onion for 2 minutes.

3 Drain the artichokes well and add to the pan, turning the heat down to its lowest. Cover and cook for about 5 minutes.

4 Pour in the stock and simmer gently for another 10 minutes. Allow to cool a little and blend to a purée.

A

5 If using thawed frozen purée, fry the onions until completely soft, stir in the purée and about 300 ml/½ pint stock, adding more if necessary. Simmer for 10 minutes and allow to cool. Process in a blender to make a thinnish purée.

6 Melt the butter in the pan and whisk in the flour with a balloon whisk, followed by the milk, whisking all the time until the mixture is thick.

7 Stir in the artichoke purée, season with black pepper. Only add salt if you think it needs it, as the vegetable stock may be salty.

8 Serve garnished with fresh parsley.

To freeze: Pour into a freezer bag or rigid container when cold and freeze for up to 3 months. To use, reheat slowly from frozen.

ASPARAGUS ❄ ◻

Spring has finally arrived when the first young shoots of asparagus start pushing their way through the soil of your asparagus bed. Although asparagus takes a couple of years to become established, once it gets going it can keep producing the delicious spears for between 8 and 20 years. So if you have the space in your garden it is worth considering planting some crowns as they are easy to grow and relatively unattractive to pests.

Asparagus has a short cropping season and is at its best eaten freshly harvested, but they do freeze well and can be bottled in brine, although they will taste more like tinned asparagus.

Varieties
o **Cito F1:** Produces heavy crops.
o **Connover's Colossal:** A good early cropping variety.
o **Franklim:** A heavy cropper of thick spears.
o **Gijnlim:** An excellent variety that has outperformed others in growing trials.

Cooking and preparing
This delicious shoot loses its sweetness rapidly so eat it fresh, raw or cooked, soon after cutting with a traditional home-made hollandaise sauce. Old asparagus becomes woody and dull. To prolong storage time, plunge the spears into boiling water for a few

A

seconds, then refresh in ice-cold water. Keep wrapped in a damp cloth in the fridge for a couple of days and reheat as required.

Asparagus makes a good savoury tart and the sprue, the very thin distorted spears, and any other rejects can be transformed into a delicious soup; cooked asparagus goes well in a wholesome salad of tomatoes, hard-boiled eggs, early new potatoes and ham with a thinned down mayonnaise or yoghurt dressing. You can also chargrill asparagus on a griddle or barbecue.

o **To prepare:** Wash and cut off the woody ends of the asparagus with a sharp knife.
o **To boil:** It is worth buying a special asparagus pan, which is tall and thin and has a lid, to cook the spears in as the tender tips will cook quicker than the bases and are best kept out of the water to steam. Tie up the spears in a bundle or several small bundles and stand them upright in a pan a third full of boiling, salted water. Cover and cook for 5–10 minutes depending on the thickness of the spears. Be careful not to overcook them. Lift out the bundles, untie the string and leave to drain. Keep the cooking water for soup.
o **To steam:** If you don't have a tall enough pan, wrap and seal the bundle of spears in foil to make a waterproof package and put it in a pan of boiling water to steam for 10–15 minutes. Remove from the pan, unwrap and leave to drain on kitchen paper.
o **To blanch:** Plunge the spears into boiling water, bringing it back to the boil for 2–4 minutes, depending on the thickness of the stems, then refresh them quickly in ice-cold water for the same amount of time. Drain carefully and roll in kitchen paper to remove excess water.

Freezing ❄

Asparagus must be treated carefully when frozen as the tender stems break easily, so it is best to store them horizontal in rigid plastic boxes rather than freezer bags.

o **To freeze:** Blanch them first as above and, when completely dry and cold, pack them in rigid containers and freeze.
o **To thaw and use:** Cook the frozen spears for 7 minutes in boiling, salted water. Drain well and use as fresh.
o **Storage time:** 12 months.

A

Bottling 🔲

Asparagus looks great bottled in tall, thin jars. Cut off the bases of the spears so that they are equal in length and width and come up to within 2.5 cm/1 in of the top of the jars you are using. Blanch the spears as above and bottle in brine (page 41).

──Asparagus Tart──────────

Serves 4–5

About 20 asparagus spears

3 large eggs

150 ml/5fl oz double cream

90 ml/3fl oz whole milk

2.5 ml/½ tsp ground coriander

Salt to taste

225 g/8 oz shortcrust pastry

20 cm/8in flan tin, or foil tin for freezing, 4 cm/1½ in deep, greased

1   Preheat the oven to 180°C/350°F/gas 4.

2   Stack the asparagus upright in a pan quarter-full of boiling water and steam for 5–10 minutes until tender. Cool and cut into 5 cm/ 2 in pieces, discarding any tough stems.

3   In a bowl, whisk the eggs lightly. Gradually whisk in the cream, milk coriander and salt.

4   Line the tin with the pastry. Place the asparagus on top and pour over the cream mixture.

5   Bake in the oven for about 45 minutes until lightly browned.

To freeze: Allow to cool completely, wrap in foil and place in a freezer bag. Remove the air, seal, label and store for up to 6 months. To use, thaw overnight in the fridge and reheat in a preheated oven at 180°C/350°F/gas 4 for 15–20 minutes until piping hot.

A AUBERGINES ❄

The aubergine is a popular plant to grow in a greenhouse. It can do well in a sheltered plot, but it does need a long, relatively hot summer to produce a decent crop. Some varieties can be grown in pots on a south-facing patio. Pick them to eat when they are young and shiny. The best method of storage is freezing them after they have been cooked in a delicious Mediterranean dish.

Varieties

o **Baby Rosanna:** Compact plant that is perfect for the patio, producing small, golf-ball sized black fruits throughout the summer.
o **Calliope:** A variety that does not need to be doused in salt to draw out bitter juices.
o **Moneymaker:** A classic, dark purple-fruited aubergine.

Cooking and preparation

These attractive but unusual looking veg are delicious in a variety of dishes from the classic Provençal dish ratatouille to the traditional Greek moussaka. They can be roasted in the oven with a medley of vegetables, such as peppers, tomatoes, courgettes and onions, or chargrilled on a barbecue. If halved, brushed with olive oil and baked in the oven in foil, you can scrape out the flesh, mix it with herbs, crushed garlic and spices to create a dip or pâté.

Aubergines soak up the juices and oils of whatever they are being cooked in, so although today's newer varieties may no longer have bitter juices to be salted out of them, it is still a good idea to slice them and layer them with salt in a colander for 30 minutes then wash, drain and dry them before frying to prevent them from absorbing too much oil. They grill well brushed with olive oil.

o **To prepare:** Wipe them clean, cut off the stalk with a stainless steel knife and be ready to brush the flesh with lemon juice to prevent discoloration after slicing.

Freezing

As they are made up of 90 per cent water, aubergines lose a lot of their texture if frozen on their own after blanching, so it is better to freeze them already cooked in dishes.

Baba Ghanoush

Serves 6

3 medium-sized aubergines
Salt
10 ml/2 tsp ground coriander
30 ml/2 tbsp olive oil
Juice of 1 lemon
30 ml/2 tbsp tahini
5 ml/1 tsp tomato purée
1 clove garlic, crushed
Warmed pitta bread to serve

1 Preheat the oven to 200°C/400°F/gas 6. Wash, dry and halve the aubergines lengthways, make a criss-cross pattern of cuts with a sharp knife, being careful not to cut through the skin, and place on a baking sheet.

2 Sprinkle the halves with salt, coriander and olive oil and bake in the oven for 20–25 minutes until the flesh is soft. Allow to cool.

3 Scoop the flesh into a food processor and blend with the remaining ingredients. Serve with strips of warmed pitta bread as a dip or as a side dish to a main meal.

To freeze: You can either freeze the aubergine halves after Step 2 in rigid containers or pack the purée in freezer bags or rigid containers in usable quantities. Store for 6 months. To use, thaw in the fridge overnight, give a quick whiz in the processor if necessary and heat through gently, if desired.

B BEANS, FRENCH ☀ ❄ ▤

French beans are easy to grow and there is such an assortment of types to choose from – dwarf, climbing, flat-podded, purple-podded, yellow, speckled red – all with different coloured flowers that will brighten up the vegetable garden or the flower borders. Their delicious flavour and the abundance of their crops make them a popular choice for the garden. The seeds of some varieties dry well, otherwise it is best to freeze any surplus. French beans are often added to a pickle of mixed vegetables.

Varieties

If you haven't got much space and you want a good surplus of beans, grow the climbing variety. They are also not such a temptation for slugs as they are not all dangling a few millimetres off the ground.

- o **Cobra:** A heavy yielding climbing bean with round green pods about 18 cm/7 in long; mauve flowers.
- o **Goldfield:** A yellow flat-podded climbing bean with flavoursome pods that can grow up to 25 cm/10 in long.
- o **Blauhilde:** Climbing purple-podded bean with tasty pods growing in bunches to lengths of up 27 cm/11 in.
- o **Borlotto:** Available as dwarf beans and climbing (Firetongue), these attractive red-streaked beans can be eaten young, whole – the red changes to green during cooking; the seeds can be eaten fresh, when they are just ripe, or dried (haricots) when they are larger and fully matured.
- o **Delinel:** Fine flavoured, high-yielding, green dwarf bean that has dark seeds. The more you pick, the more you get!

Cooking and preparing

Eat French beans as soon after picking as you can when they are young and crisp and full of flavour. At this stage you can eat them raw or just cooked in salads, or steamed and served with butter and mixed with savory as a side dish. When beans are a little past their best, boil them until just cooked, drain and sauté them in butter, crushed garlic and a squirt of tomato purée for a couple of minutes – no longer as they will begin to fall apart.

The Borlotto variety, which is more Italian than French and known generally as the borlotti bean, can be eaten young in its pod, or when the seeds have nearly matured, pod them and boil until just tender, about 5 minutes, drain, and stir in some olive oil, crushed garlic (or garlic oil) and savory or thyme.

o **To prepare:** Pinch off the top stem part of the pod, or snip a handful off at once with scissors; there is no need to cut off the pointed ends. The flat-podded beans may need stringing (pages 88–89). If you pick any variety of bean when they are quite long, chop them into 5 cm/2 in pieces.

o **To cook:** Either steam or boil in lightly salted water. There are two schools of thought on how long beans should be cooked: many prefer them crunchy, blanched in boiling water for 3 minutes; others find that they are more flavoursome if cooked for longer until tender, 7–10 minutes. Whatever you do, don't overcook them.

Drying ☼

Leave the red-streaked Borlotto variety of bean to dry on the plant until the pods have shrivelled and gone brown and the seeds are plump inside. Remove the seeds from the pods and discard any that are discoloured. If the seeds have dried completely, store immediately in dry jars in a cool, dry place; if they haven't, lay them out on a tray and finish the process in a warm, airy place.

Freezing ❄

French beans freeze well, retaining their flavour and texture better than runner beans (page 88). Only the freshest, best quality beans should be frozen.

o **To prepare:** Trim off the stalk ends and either leave whole or chop into smaller pieces if long. Blanch for 2–3 minutes.

o **To freeze:** When cold, place into freezer bags or rigid containers and freeze.

o **To thaw and use:** Cook from frozen in boiling, lightly salted water for 6–8 minutes or until just tender.

o **Storage time:** 12 months.

B ## Preserving 🖾

French beans can be added to a mixed vegetable pickle, such as piccalilli (page 114) or they can be pickled on their own in spiced vinegar. Either leave them whole, topped and tailed, or cut them into 2.5 cm/1 in pieces. Boil for 5 minutes in lightly salted water. Drain well before standing whole ones upright in jars, packed tight, or packing in the chopped up ones. Pickle with wine or cider vinegar, dill seeds and whole allspice berries.

BEANS, RUNNER ❄ 🖾

Runner beans are very prolific in the right conditions. They do not like it too hot or dry, so in a hot summer it is important to make sure that they are well watered and mulched in order to keep their roots moist. A good runner bean crop will need picking every day as if they are left on the plant for too long they will grow tough and stringy. They are also best eaten within two days, as the beans' sugars turn rapidly to starch soon after picking. They can be frozen and can be added to a spicy chutney or pickle in the same way as French beans (page 86).

Varieties

o **Achievement Merit:** A good runner for freezing, producing an abundance of long straight pods; scarlet flowers.

o **Enorma:** Another good freezer and a popular high yielding variety with scarlet flowers and long pods.

o **White Lady:** Has white flowers that are less attractive to birds and can withstand hot summers.

Cooking and preparing

An ideal vegetable side dish that will go with everything – meat, fish and on their own in a cheesy sauce, or a bean salad, or just smothered in butter or for the more health conscious, lemon or lime juice, with lashings of freshly ground black pepper and a sprinkling of sea salt.

o **To prepare:** Pick the beans young to encourage more to grow, and they are less likely to need stringing. Leave them whole, just snipping off the stalk end and steam, or slice diagonally down the bean. Larger beans will need stringing:

with a sharp knife start by cutting off the stalk end and continue running the knife down each side, being careful not to cut into the flesh of the bean.

o **To cook:** Sliced runner beans can be stir-fried, steamed or boiled until tender but not crunchy for maximum flavour.

Freezing ❄
Only pick the beans when you are ready to freeze them and have the water boiling ready to blanch them, so that no goodness is lost in the waiting.

o **To prepare:** If the beans are picked young you only need to cut off the stalk end and snap them or cut them into shorter lengths, if long, before plunging them into the boiling water and blanching them for 2 minutes. Older beans will need stringing, chopping in 5 cm/2 in chunks (not slicing), to maintain flavour, and blanching for a minute longer.

o **To freeze:** Make sure the beans are properly dry before sealing them in freezer bags or rigid containers and freezing.

o **To thaw and use:** Boil from frozen for 7–10 minutes or until tender.

o **Storage time:** 12 months.

Preserving ▨
Runner beans make a delicious chutney and are a good way of using up a glut.

── Classic Bean Chutney ─────────
Makes about 900 g/2 lb

700 g/1½ lb demerara sugar

900 ml/1½ pints malt vinegar

900 g/2 lb runner beans

450 g/1 lb onions, finely chopped

1 garlic clove, crushed

5–15 ml/1–3 tsp spices, such as cardamom seeds and fenugreek

15 ml/1 tbsp spices, such as cumin, turmeric and powdered English mustard

1 Dissolve the sugar in the vinegar.

2 Add the beans, onions, garlic and spices and boil for 15 minutes.

3 Simmer for a further 15 minutes or until the mixture starts to thicken, stirring occasionally.

4 Allow to cool a little and spoon into warmed, sterilised jars. Seal and label.

5 Store in a cool dark place for a few months before using.

BEETROOT

Beetroot come in different shapes and colours and are straightforward to grow, but make sure you grow varieties known for their flavour so that none is lost in the storage. It is also a very versatile crop to keep; traditionally it makes a good pickle, but it can also be turned into a delicious wine and a healthy juice. The roots can be frozen when cooked, and the leaves can be treated like spinach (page 188).

Varieties

o **Cheltenham Green Top:** A cylindrical variety, full of flavour, that stores well.

o **Boltardy:** The most popular ball-shaped type because – as its name suggests – it is not prone to bolting (running to seed). It is also very tasty.

Cooking and preparing

Beetroot are traditionally eaten raw or cooked in salads but make delicious soups, too – thick or thin, as in the traditional Eastern European Borscht. The roots can be used in baking, too, giving a good flavour and a rich, dark colour to bread, chocolate cake and muffins, which can then be frozen. They also go well in savoury bakes, although the dish will look a little pink.

o **To prepare:** Wash and cut off the leaves, leaving about a 12 mm/½ in stalk. Be careful not to cut into the skin as the juice will bleed and lose flavour during cooking.

o **To boil:** Cook whole in boiling water, for about 20–30 minutes until tender or use a pressure cooker. Drain, rinse in cold water, drain again and dry on kitchen paper. Peel off the skin and cut off the top.

o **To roast:** Place in a baking dish, brush with olive oil and bake in the oven at 180°C/350°F/gas 4 for about an hour until tender. Peel off the skin and cut off the top.

Beetroot Muffins

Makes 12

200 g/7 oz plain flour
25 g/1 oz cocoa powder
15 ml/1 tbsp baking powder
100 g/4 oz caster sugar
50 g/2 oz butter, softened
5 ml/1 tsp vanilla essence
2 eggs, beaten
120 ml/4 fl oz milk
100 g/4 oz raw beetroot, grated
100 g/4 oz chocolate drops

1   Preheat the oven to 180°C/350°F/gas 4. Grease or put paper cases in muffin tins.
2   Mix together the dry ingredients.
3   Stir in the remaining ingredients until just blended.
4   Spoon into the tins or paper cases and bake for about 20 minutes until just firm.

To freeze: When cool, open freeze and pack into rigid containers, interleaving with greaseproof paper or baking parchment. Store for up to 3 months. To use, thaw at room temperature for 2 hours.

Storing

Lift the roots from the ground, cut off their leaves to about 1 cm/ ½ in from the root to avoid bleeding, and wash and dry them well. Layer the roots in boxes of sand or dry compost and store them in a shed or garage. Alternatively, you can store them in a clamp (page 18).

B Freezing ❄

Beetroot freezes well but is best if cooked first.

o **To prepare:** Cook the beetroot, remove the skins and cut into chunks (as above).
o **To freeze:** Freeze in rigid containers or freezer bags.
o **To thaw and use:** Defrost and use as fresh.
o **Storage time:** 6 months.

Juicing ◯

As beetroot has a strong rich flavour, you may prefer to blend it with another vegetable or fruit juice, such as carrot or apple. Store in the freezer.

Preserving ▨

With its rich red colour and distinctive sweetness, beetroot has been pickled for many a century, making a tasty accompaniment to cold meats and cheeses. Choose small beets for best results.

o **To prepare:** Clean and cook until tender, either boiling or roasting, then place whole if small or cut into small chunks, in a sterilised jar, filling it about three-quarters full.
o **To pickle:** Fill the jar with malt or wine vinegar and 5 ml/ 1 tsp pickling spices.
o **Storage time:** Several years.

Beetroot can also be made into a tasty chutney, follow the Traditional Apple Chutney recipe (page 72), replacing the apples with beetroot.

Wine making ⧗

The sweet earthy taste of beetroot translates into an unusual but delicious wine. To prepare, scrub the roots well before dicing them. Bring them to the boil and simmer gently for an hour. Follow the directions on pages 62–65, adding 225 g/8 oz chopped raisins to the fermentation bucket where directed.

BLACKBERRIES ❄ ◐ ▨ ♟   B

From late summer the hedgerows are laden with blackberries and you will notice that there are many different types, ranging from small, tightly packed sour berries to the larger, sweet and juicy ones. Growing your own isn't such a mad idea either, as some varieties of cultivated blackberries are thornless and produce large flavoursome, plump fruits suitable for making preserves, wine, juice and for freezing. They also grow into a compact plant, so there are no runners shooting themselves around the garden. Hybrids include loganberries and boysenberries, which can be grown and used in the same way.

Varieties

o **Loch Ness:** A high-yielding, relatively compact, thornless blackberry only growing as tall as 1.8 m/6 ft – some can reach up to 4.5 m/15 ft.

o **Merton Thornless:** Another compact variety that produces a good crop of tasty, medium-sized berries.

o **Loganberry:** A cross between a blackberry and a raspberry producing long, dark red berries with a strong, tart flavour.

o **Boysenberry:** A cross between a blackberry, a raspberry and a loganberry, this hybrid produces reddish purple fruits that taste more like a wild blackberry.

Cooking and preparing

Traditionally paired with apples in pies, crumbles, jams, jellies and juices, blackberries, whether wild or cultivated, enhance every dish they are added to. Their intense flavour lends itself to producing a delicious wine and providing a sweet note to rich meat dishes such as duck and game. Plump berries tint puddings, cakes and baked cheesecakes with tantalising purple juice and autumnal sweetness. They also bake well on top of a tart filled with crème fraîche, make a delicious late summer pudding and, fresh from the plant, make a tasty topping for a late summer pavlova.

o **To prepare:** Pick over the fruit, discarding stalks and unripe or overripe berries. Only wash if necessary and allow to dry thoroughly.

B Freezing ❄

Freezing is the simplest and most useful method of storing blackberries as they can be open frozen straight away after picking and used straight from the freezer whenever needed. The fruits do not tend to lose their shape or substance when defrosted, so can be used raw in fruit salads or on top of pavlovas.

- o **To freeze:** Prepare as above and place on a baking tray in one layer. When frozen, pack into rigid containers or freezer bags being careful to remove any air from the bags.
- o **To thaw and use:** Defrost at room temperature or in the fridge overnight and use as fresh.
- o **Storage time:** 12 months.

Juicing ◌

To make blackberry juice it will certainly be easier with an electric juicer because of all the pips. Alternatively, process the fruit in a blender and sieve it to remove the pips. Choose well-ripened fruit that are not over ripe. Blackberry juice blends well with other fruit juices, such as apple, raspberry and blueberry. Store in the freezer.

Preserving ▨

Blackberries can be made into jam on their own or using half and half with other soft fruits or peeled, cored and chopped apples. For a pip-free jam, rub through a sieve before adding the sugar.

Blackberry and Apple Jam

Makes about 1.5 kg/3 lb

450 g/1 lb apples
300 ml/½ pint water
450 g/1 lb blackberries
1 kg/2¼ lb sugar

1   Peel and core the apples and cut them into pieces. Put into a pan with the water and boil to a soft purée.

2   Prepare the blackberries, add them to the apples and bring back to the boil. Cook until the berries are just soft.

3   Stir in the warmed sugar until dissolved. Boil hard to setting point.

4   Stir well put into warmed, sterilised jars, seal and label.

Make a jelly if you prefer to eliminate the pips completely. You will need to give pectin levels a boost with fresh lemon juice if you are only using blackberries. For 1.75 kg/4 lb blackberries, add the juice of 2 lemons and use 300 ml/½ pint water and follow the procedure on page 48.

• **Storage time:** About 12 months unopened in a cool, dark place; once opened, it will keep in the fridge for about a month.

Spicy Blackberry Chutney

This is a treat to accompany cheese and cold meats and will be ready for Christmas. For 2 kg/4½ lb blackberries use 1 kg/2¼ lb sugar, 450 g/1 lb sliced red onions, 600 ml/1 pint vinegar, 10 ml/2 tsp each ground cloves and cinnamon, 5 ml/1 tsp ground allspice and 45 ml/ 3 tbsp chopped fresh ginger, peeled. Follow the recipe on page 54.

• **Storage time:** Several years unopened in a cool dark place; once opened, up to 12 months also in a cool, dark place.

Bramble Ketchup

Simply purée the Spicy Blackberry Chutney and push it through a sieve, then either freeze it in usable quantities in rigid containers. When defrosted, keep it in the fridge for up to 2 months. Alternatively, pour it into sterilised jars or bottles.

Wine making 🍷

Blackberries make a delicious, rounded-flavoured country wine. For 2 kg/4½ lb prepared, well-washed fruit, you need 450 g/1 lb chopped raisins, 4.5 litres/1 gallon cold water, 5 ml/1 tsp pectinase, 2 Campden tablets and a yeast starter bottle.

1 Mash the berries with a wooden spoon in a fermentation bucket, then add the raisins.
2 Pour the cold water over the fruit, stir well and add pectinase, the pectin-destroying enzyme and 1 Campden tablet.
3 Cover the bin and leave for 24 hours.
4 Continue from Step 3 on page 63.

B

Blackberry Liqueur

Place 500 g/18 oz blackberries, as dry as possible, with 250 g/9 oz sugar in a large covered jar with 1 litre/1¾ pints vodka, gin or whisky. Shake to dissolve the sugar, then cover. Leave in a cool place away from sunlight and shake 2–3 times a day for 2 weeks, then once a week for 6 weeks. Strain through a muslin and pour into bottles.

BLUEBERRIES ❄ ◗ ▢ ▨

Since the start of the 21st century the cultivated blueberry, as opposed to the wild, much smaller bilberry native to the British Isles, has become mainstream in British gardens and kitchens. The first fruit to be classed as a superfood for its health-enhancing properties, the traditional Native American fruit is now grown easily in the UK, especially in pots on the patio.

Blueberries are self-fertile but will produce heavier yields if there is more than one plant; they need acidic moist soil to flourish. They are wonderful fresh in muesli or baked in cakes, and store well in the fridge and freezer. Like blackcurrants, blueberries make a delicious jam and combine well with other fruit for juicing.

Varieties
o **Duke:** Medium to large, light blue berries on a plant that matures at a height of 1.2–2 m/5–6 ft 6 in. Heavy cropping, eventually giving a yield of 9 kg/20 lb.
o **Bluecrop:** The most popular variety, with a heavy crop of pale blue, flavoursome berries.
o **Coville:** Produces high yields of sweet berries.

Cooking and preparing
Blueberries add a burst of sweetness and a pleasant purple hue to many recipes. Use them to make jams, pies, tarts, ice creams and cheesecakes. They also provide wonderful splashes of colour and flavour to muffins and pancakes. Eat the fruit raw in cereals, fruit salads and mixed with raspberries and strawberries on top of pavlovas, or lightly poach them in a little water, sugar and lemon juice and eat with ice cream or pannacotta.

B

o **To prepare:** Blueberries grow in bunches and if you run these gently through your fingers when some of the berries appear ripe (completely blue), the ripe ones will come away, leaving those that need a little longer on the plant. After picking leave them in a container with the lid off for a few hours before covering and placing in the fridge unwashed. This prevents a build-up of condensation, which can cause the fruit to deteriorate. Don't wash them until you are about to use them.

o **To store:** The fruit will keep in the fridge for up to a week if stored in a single layer.

Freezing ❄

Select just-ripe fruit with firm flesh. Avoid any berries whose juice is running.

o **To prepare:** Pick over, wash and dry on kitchen paper.

o **To freeze:** Open freeze, pour into freezer bags and return to the freezer. Alternatively, pack in rigid containers with 125 g/ 4 oz caster sugar to 450 g/1 lb blueberries.

o **To thaw and use:** Thaw at room temperature for 3 hours or overnight in the fridge. Open frozen blueberries may lose their shape when defrosted, but still look attractive in fruit salads and other fresh fruit desserts.

o **Storage time:** 3 months when open frozen; 12 months in sugar.

Juicing ◌

Riper fruit can be used for juicing, but be sure to discard any mouldy berries. Try a combination of blueberries and raspberries. Store in the freezer.

Bottling ▣

The fine flavour of blueberries will keep for years when bottled in syrup or in rum. Use on ice cream, sorbets, cheesecake and mousses.

B Preserving 🖳

Blueberries make a rich and flavoursome jam, delicious on croissants for breakfast, or layered with cream in a Victoria sponge cake. As the skins toughen when cooked with sugar, boil the fruit in water until the berries are soft and the juice is running before adding the sugar.

o **To make jam:** For 2.25 kg/5 lb jam, simmer 900 g/2 lb blueberries in 600 ml/1 pint water for 10–15 minutes or until soft. Add 1.25 kg/2½ lb sugar, bring slowly to the boil and follow the instructions on page 48 from Step 5.

——— Blueberry Pie ———

This is a variation on an American theme. You can use blueberries straight from the freezer, or bottled blueberries, drained, as well as fresh. You can also freeze the pastry before cooking it, so that you have the components for a delicious dessert at hand when you need them.
Serves 6

For the pastry:
175 g/6 oz plain flour
100 g/4 oz polenta
30 ml/2 tbsp caster sugar
2.5 ml/½ tsp salt
200 g/7 oz cold unsalted butter, cut into small pieces
15–45 ml/1–3 tbsp iced water to bind
For the filling:
45 ml/3 tbsp caster sugar
220 ml/small tub crème fraîche
450 g/1 lb blueberries

1 Pulse all the pastry ingredients, except for the water, in a food processor until it resembles breadcrumbs. Stir in enough water to form a dough.

2 Flatten it into a disc shape, wrap it in clingfilm and put it in the fridge for 30 minutes. Alternatively you can freeze it at this stage, wrapped again in a freezer bag. To use, thaw it in the fridge overnight.

3 Preheat the oven to 190°C/375°F/gas 5 and line a large baking sheet with lightly oiled greaseproof paper.

4 Roll the pastry out into a circle about 25 cm/10 in diameter and place it on the baking sheet. Dampen the edge of the pastry with some water and roll it over to create a rough, thickish rim of about 2.5 cm/1 in high, leaving an inside diameter of roughly 20 cm/8in.

5 Mix half the sugar into the crème fraîche and dollop it over the pastry base, spreading it lightly as you go. Scatter the blueberries over the top and sprinkle with the rest of the sugar.

6 Bake in the oven for 20–30 minutes or until the pastry is cooked and golden.

BROAD BEANS ❄ 🍷

A great leafy plant that can be sown in the autumn or early winter to produce long pods of delicious beans to eat when there is not much else in the garden. Picked young, the whole pods can be eaten as well as the growing tips. The beans freeze well without losing any of their delicate flavour and they also make an interesting wine.

Varieties

o **Aquadulce Claudia:** Best variety for early or late sowings for an early crop of beans with white seeds.

o **Bunyard's Exhibition:** An old variety that grows up to 1.2 m/4 ft tall, producing good crops in all sorts of soil.

o **Imperial Green Longpod:** Another tall one that produces long pods, as the name suggests, containing up to 9 green beans in each.

Cooking and preparing

Steam young beans in their pods and throw in the growing tips as well and serve with garlic, butter and fresh mint leaves stirred in. By removing the growing tips, the plant will put more energy into creating longer and more numerous beans and will also prevent the dreaded blackfly from moving in. As they grow bigger, the pods become tougher and cannot be eaten. However the beans inside, whether white or green, will grow plumper and when cooked make a delicious addition to salads, or can be mixed with herbs, such as mint, thyme or savory, for a simple side dish. Older beans can be skinned, boiled and puréed to accompany a dish of lamb.

B

o **To prepare:** Wash, top and tail young pods and steam whole. Remove older, larger beans from their tougher pods and boil. Skin individual beans that are past their best.

o **To skin:** Before skinning individual bean seeds, blanch them in boiling water for 1 minute, drain and place on a clean tea towel and slip the beans out of their skins.

o **To boil:** Do not cook beans in salted water as the salt toughens their skins. Cook them in boiling water for 10–15 minutes, drain and then season to taste. If the beans are a little on the old side, remove the skins from the individual beans as above before boiling them in slightly salted water for 10–15 minutes.

Freezing ❄

Once picked, don't keep fresh beans too long as they will start to lose their flavour.

o **To prepare:** Remove the beans from the pods and separate into small and larger sizes. Blanch the small ones for 2 minutes and the larger ones for 3 minutes.

o **To freeze:** Pack into freezer bags or rigid containers and freeze.

o **To thaw and use:** Cook from frozen in boiling water for 6–8 minutes, depending on the size, or until tender.

o **Storage time:** 12 months.

Wine making 🍷

Not many people can say that they've been offered a glass of broad bean wine to drink.

1 Remove the beans from their pods to make a quantity of 2 kg/4½ lb and bring them to the boil in 4.5 litres/1 gallon water with the grated rind of a lemon. Simmer for an hour.

2 Cover and leave to cool, then strain the liquid on to 225 g/ 8 oz chopped raisins, lemon juice, 5 ml/1 tsp tannin, a sachet of wine yeast and 5 ml/1 tsp yeast nutrient in a fermentation bucket.

3 Cover well and leave to ferment for 4–5 days, stirring daily.

4 Strain the liquid off through a fine mesh bag and stir in 1.1 kg/2½ lb granulated sugar, making sure it dissolves.

5 Continue from Step 8 on page 63. Before racking the wine, a hydrometer reading should be around 1.000.

6 Once bottled, mature for 6–9 months before drinking.

BROCCOLI ◨ ❄

There are three types of this hardy, highly nutritious vegetable: calabrese, the large green-headed broccoli; purple-sprouting broccoli on thinner stems; and the less common white sprouting variety. If you plant wisely, you can be harvesting broccoli almost all year round.

Varieties

o **Calabrese Marathon:** A popular variety that produces one large bluish-green head in late summer/early autumn and after cutting several smaller ones will appear. Sow in the spring.

o **Bordeaux:** A purple sprouting variety that can be harvested as early as July, if sown in February.

o **Claret:** Large purple sprouting heads with thick stems that can be harvested from March to April.

o **Early Purple Sprouting:** Produces spears from February to May.

o **Extra Early Rudolph:** A very early purple sprouting variety, ready to harvest from January.

o **Early White Sprouting:** Tall white cauliflower-like heads from February.

Cooking and preparing

Eat broccoli straight from the garden, lightly boiled or steamed to retain as much of its superfood goodness as possible. With sprouting broccoli, cut and cook the flower stems and the smaller leaves as well. The younger the broccoli, the shorter the cooking time. Parboil the plumper calabrese florets and toss them in olive oil with crushed anchovy, chillies and garlic, then stir them into pasta or risotto, or add florets to stir-fries for texture, colour and flavour. Add broccoli to cauliflower cheese with sliced tomatoes and freeze it, wrapped in foil and in a freezer bag (stores for 3 months), or serve it lightly cooked with a vinaigrette.

B

o **To prepare:** Remove, dry or discoloured leaves and trim off the fibrous part of the stem base.

o **To cook:** Plunge calabrese into boiling, salted water and boil or steam for 5 minutes. Avoid overcooking as it will lose its fresh green colour. As the flower heads of the purple sprouting varieties are more delicate and go mushy if overcooked, cook them in the same way as you would asparagus (page 82), standing the stalks in the boiling water, so that the heads cook in the steam. Alternatively, wrap the spears in foil with some butter and seasoning, making the parcel watertight, and boil it in water for 5–8 minutes.

Storing 🖼️

Purple sprouting broccoli keeps growing through the winter and can start sprouting as early as January, so leave it in the ground and pick enough spears for a meal when you want them, which will encourage new spears to grow.

Freezing ❄️

As the flower heads of purple sprouting broccoli can overcook so easily, it is best to just harvest it from the garden when you need some, but calabrese freezes very well.

o **To prepare:** Select firm, fresh, bright green heads. Wash and divide them into small, even-sized florets.

o **To freeze:** Blanch for 2 minutes and open freeze. Pack in freezer bags and return to freezer.

o **To thaw and use:** Cook from frozen in lightly salted boiling water for 7–8 minutes.

o **Storage time:** 12 months.

BRUSSELS SPROUTS 🔳 ❄️ **B**

Brussels sprouts have endured a lot of bad press, probably because they are so often overcooked. Lightly boiled or steamed fresh from the garden, the young sprouts are sweet and nutritious. Although they take up a lot of space on the allotment or in the vegetable plot, they keep going all through the winter and you can leave them on the stalks until you are ready for them. The top of the plant is delicious, too.

Varieties

When the plants are young in the summer, cover them with horticultural fleece to keep pigeons and the cabbage white butterfly at bay.

- o **Brilliant F1:** An early variety with a long harvesting period. Disease resistant and not prone to bolting.
- o **Cascade:** Good in all weathers and resistant to mildew.
- o **Red Rubine:** A red Brussels sprout that keeps its colour during cooking.
- o **Trafalgar F1:** A heavy cropping variety that produces firm sweet sprouts.

Cooking and preparing

Brussels sprouts are at their best lightly boiled or steamed then sautéed for a couple of minutes with peeled, chopped and cooked chestnuts, or toasted pine nuts, just before serving. Add some crisply fried strips of pancetta or streaky bacon to make it more of a dish. Alternatively, sprouts can be shredded and stir-fried. For an interesting side dish, throw in some thinly sliced leeks too.

Brussels tops can be treated like spring greens: wash, slice and boil in a small amount of salted water for about 5 minutes until tender. Serve immediately with a squeeze of lemon juice.

- o **To prepare:** Use the sprouts as soon as you have picked them off their stems, to retain maximum flavour. Peel back and discard the outer, discoloured leaves and rinse well.
- o **To cook:** To bring out the best in your sprouts, boil small firm buttons in deep salty water for 5–8 minutes, or steam them, until just tender. Drain and refresh under the cold tap to prevent them from cooking any further if you're not going to eat them straight away. Before serving, return the sprouts

B

to the pan, add a knob of butter with plenty of freshly
ground pepper and gently reheat.

Storing 📼

Brussels sprouts can be left in the garden throughout the winter. In
fact, they taste much better once they have been frosted. They can
also be cut and stored for several weeks in a cool, dark place in a
rodent-free shed, for example, or larder, as long as they are left on
the stalks with the leaves removed.

Freezing ❄

You can freeze sprouts, but they do lose some of their texture when
defrosted.

- **To freeze:** Prepare as above and sort them into small and
  large sizes. Blanch the small ones for 1½ minutes and the
  larger ones for 3 minutes. Open freeze, tip into freezer bags
  and return to the freezer.
- **To thaw and use:** Cook from frozen in slightly salted water
  for 5–8 minutes according to their size, until just tender.
- **Storage time:** 12 months.

CABBAGE 🖻 ❄ 🌢 🗏

C

There is a different type of cabbage for every season so you can eat fresh cabbage all year round if you have the space to grow it. They are relatively easy to grow but require attention to keep pests at bay. The red and white types of cabbage are best for storing and pickling. Contrary to popular belief, you can freeze raw cabbage with good results.

Varieties

o **Primero F1:** A red cabbage that produces firm, compact heads with little core and keeps well in the ground.

o **Kilaxy F1:** Holland winter white cabbage that matures in autumn and stores well.

o **Kilaton F1:** A smooth ball-head cabbage similar to the Holland white that matures in November and keeps well in the ground and stores for a long period once cut. Resistant to clubroot.

o **Resolution F1:** Savoy cabbage with dark green head that will remain in good condition on the plot once it reaches maturity. Harvest from mid-winter to late spring.

o **Advantage F1:** A loose-leafed cabbage that can be planted in late summer to produce an early crop of spring greens.

o **Hispi F1:** A pointed loose-leafed cabbage that is hardy and can be sown from March to October to produce a crop all year round.

o **January King 3:** A hardy winter variety with a great flavour and very reliable. Keeps well in the ground.

Cooking and preparing

Cabbage has a bad reputation, especially with children, due to it being overcooked. Packed with nutrients, cabbage is not just a vegetable for boiling but the red and white varieties can be eaten raw in salads and shredded and mixed with mayonnaise and grated carrot to make coleslaw. They can also be pickled, with white cabbage being the main ingredient for sauerkraut. Red cabbage rapidly loses its colour and flavour during cooking, so add a little wine, vinegar or apples, or all of them, to the pan. The dark-leafed Savoy and January King are delicious lightly boiled and served with a knob of butter and a scattering of caraway seeds stirred in.

C

o **To prepare:** Pull off the shabby outer leaves but don't discard them if they are undamaged, as the darker the leaves the more nutrients they contain. Wash well or soak in salted water first to eject any insects, then rinse. Cut in quarters and with a sharp knife cut out the core if it is tough and fibrous, otherwise this part is tasty, too. Shred or roughly chop the rest.

o **To boil:** Pack the chopped cabbage into a large saucepan and sprinkle with salt. Turn on the heat to high to warm up the bottom of the pan. Pour boiling water straight from the kettle over the top to halfway up. It quickly returns to the boil. Cook fast for 5 minutes until just soft to the bite. Drain well, squeezing as much moisture out as you can. Serve straight away in a hot dish.

Storing 🔲

Most cabbages – red, white and winter varieties – can be left in the ground during the winter months. Pack straw around them if it is too wet, or if the temperature is going to drop below -10°C/14°F. They can also be stored in a cool, dark place, as long as it is rodent proof. If you are going to do this, pull the cabbages up by their roots, remove any damaged leaves and tie them up so that they hang upside down. This is important as cabbages make their own moisture and need a good airflow to keep them dry and prevent odours. Crisp, firm, tight-headed summer cabbages can also be stored in this way until February.

Freezing ❄

Freezing cabbage is not a common practice, but it can freeze well if you choose very fresh specimens with crisp, firm leaves.

o **To prepare:** After preparing and washing as above, pat dry with kitchen paper. Finely shred hard white and red varieties, cut green varieties into wider strips. If making a cabbage dish, cook it and cool it quickly.

o **To freeze:** Blanch the raw red and white hard varieties for 1 minute, for green 1½ minutes. Drain well and dry. Pack into freezer bags.

o **To thaw and use:** Boil all varieties from frozen, until just tender or use in stir-fries. Frozen white cabbage will be too limp when defrosted to use in coleslaw.

o **Storage time:** Red and white 12 months, green 6 months.

C

Juicing 〔◇〕

Cabbage can be made into a healthy drink, but is tastier blended with carrot or apple juice.

Preserving 〔▤〕

Shredded hard red and white cabbages pickle very well, using cloves, coriander seeds, root ginger and peppercorns to spice the vinegar. Put the prepared, finely shredded cabbage in a bowl, sprinkle with salt and leave overnight, then drain, rinse and dry well before pickling. About 700 g/1½ lb cabbage will fill around 5 450 g/1 lb jars.

Sauerkraut

This is a fermented pickled cabbage that can be made in a fermentation bucket (page 60) or crock. Salt is used to extract the juice, which then ferments over time. The longer and slower it ferments, the better. It is delicious cooked long and slowly in a casserole containing pork, sausages, potatoes, onions, garlic, wine and apples, and spiced with caraway seeds or juniper berries.

1   Prepare firm white cabbage as above and shred it finely.
2   Layer the cabbage in a fermentation bucket or crock, sprinkling sea salt evenly across it. Juniper berries and peppercorns can also be scattered over each layer.
3   Continue filling the bucket or crock with alternate layers of cabbage, salt and seasonings, pressing it down carefully as you go to extract the liquid, until three-quarters full.
4   Cover the crock with a cloth and place a wooden disc to fit inside, then place a heavy weight on top of this in order to keep squeezing the liquid out of the cabbage. The cloth excludes air, which would contaminate the mixture.
5   After a day or two the extracted liquid begins to ferment and rises to cover the lid. A scum will form that must be removed from time to time, but leave enough liquid to cover the lid. If the cabbage has not made enough liquid, top it up with some brine (page 41).
6   Fermentation will take about 2 months depending on the temperature, but left at a temperature of around 7°C/45°F, it will take up to a year and result in an excellently flavoured

C

sauerkraut. Keep it separate from your living space, as it doesn't smell very good.

7 When you remove sauerkraut from the crock, wash the cloth and cover before replacing them.

o **To store:** Almost fill sterilised jars with the sauerkraut. It will keep in a cool, dark place for a year or more.

o **To use:** Only rinse the sauerkraut if it is too sour for your taste. Otherwise bring a pan of the quantity you need to the boil in its own liquid and simmer it for 30 minutes.

CARROTS

Carrots are very versatile vegetables to cook and to store, and they are full of vitamins and nutrients. There are so many varieties in different shapes, sizes and colours – white, purple and yellow – that mature at different times that you can have fresh carrots all year round. Generally, carrots are easy to grow but not as reliable as some vegetables, so it is worth trying a few different varieties to find which one suits your soil. Maincrop carrots can be stored in sand, in the ground or in a good old-fashioned clamp (page 18). The thin, flavoursome, early carrots are suitable for freezing.

Varieties

o **Autumn King 2:** A classic late sowing, long-rooted carrot that will remain in good condition, once mature, in the soil throughout winter.

o **Early Nantes 2:** Can be sown under cloches in March for a crop of carrots in June. Freezes well.

o **Nigel F1:** Long, cylindrical, coreless carrots that have a good flavour. Suitable for storing.

o **Rainbow F1:** A mixture of coloured maincrop carrots.

o **Resistafly F1:** A maincrop carrot that has good resistance to carrot fly.

o **Sugarsnax 54 F1:** A long, narrow maincrop variety high in betacarotene with an intensely sweet flavour. Delicious raw. Harvest them early for the freezer. They also store well.

C

Cooking and preparing

Young and crisp, carrots can be chopped into sticks and eaten raw with dips or grated into salads. Naturally high in sugars that caramelise when cooked, later carrots enhance stews and casseroles and make wonderful soups teaming up with coriander or cumin. The addition of thyme and parsley also heightens their flavour.

Glazed carrots make a delicious accompaniment to a roast: sauté chopped carrots in butter and sugar, just cover with boiling stock and cook for 10–15 minutes; Remove the lid and cook for another 5 minutes or until most of the liquid has evaporated. Serve with plenty of freshly ground black pepper and fresh chopped parsley.

Made with wholemeal flour, brown sugar and nuts, carrot cake and muffins make a healthy treat.

o **To prepare:** Slice off the top with the stalk, including any green on the root, and chop off the thin root at the other end. Home-grown carrots need only be washed not peeled, unless they are old and have been stored for a while. Slice large, older carrots into discs or sticks, but keep young, new carrots whole.

o **To cook:** In boiling salted water, sliced or chopped carrots will take 10 minutes to cook, whole will take 15–20 minutes; add 5–10 minutes if you are steaming them. Young early carrots will cook faster.

Storing

Maincrop carrots will keep in the ground until you need them, as long as it is not too wet; otherwise they will store throughout the winter months layered in boxes of sand or similar and kept in a cool dark place. After lifting from the ground, cut back the foliage to within 5 mm/¼ in of the roots and store them when dry.

C Freezing ❄

It is worth freezing early young carrots for a taste of summer all year round. Select carrots with smooth, clean skins.

o **To prepare:** Wash and prepare the carrots as above. If they are not too big freeze them whole; slice the larger ones.

o **To freeze:** Blanch the whole ones for 3 minutes, sliced for 2 minutes. Dry and pack in freezer bags.

o **To thaw and use:** Cook from frozen in boiling salted water for about 5 minutes or until tender. Alternatively, they may be added frozen to soups and stews. Once thawed the carrots will be soft so cannot be used raw.

o **Storage time:** 12 months.

Juicing ◊

Carrots make a delicious nutritious juice to drink straight away or to be frozen or bottled (pages 41–43). It can be drunk on its own or it is a good diluter with other stronger vegetable juices, such as beetroot. Don't peel carrots before juicing, washing them well instead means you will receive the maximum nutrition from the whole vegetable. Either cut them up into 5–7.5 cm/2–3 in pieces or feed the large end into the juicer first.

—— Carrot and Apple Cake ——

Using the juice and the carrot pulp left after juicing gives this cake a deliciously moist texture.

Serves 12

350 g/12 oz caster sugar

4 eggs

300 ml/½ pint vegetable oil

350 g/12 oz plain flour

10 ml/2 tsp baking powder

5 ml/1 tsp salt

10 ml/2 tsp ground cinnamon (optional)

100 g/4 oz finely chopped peeled apple

225 g/8 oz carrot pulp

150 ml/¼ pint carrot juice or 2 carrots, juiced

225 g/8 oz chopped nuts

For the icing:
75 g/3 oz cream cheese
100 g/4 oz butter, softened
450 g/1 lb icing sugar, sieved
5 ml/1 tsp vanilla essence
Pinch of salt
50 g/2 oz chopped nuts

1 Preheat the oven to 180°C/350°F/gas 4. Line a 23 x 30 x 5 cm/9 x 12 x 2 in baking tray with greaseproof paper.

2 To make the cake, beat together the sugar, eggs and oil in a bowl.

3 Mix together the flour, baking powder, salt and cinnamon, if using, and stir into the egg mixture.

4 Add the apple, carrot pulp and juice and nuts and mix well.

5 Pour the mixture into the prepared tray and bake for 45 minutes until it is golden and has shrunk away from the sides slightly.

6 Remove from the oven and leave to stand for about 10 minutes before turning out on to a cooling rack.

7 To make the icing, beat together the cream cheese and butter. Stir in the icing sugar, vanilla essence and salt. Spread over the cooled cake and scatter the chopped nuts over the top.

To freeze: Freeze the cake without the icing. Wrap it in a double layer of clingfilm and then wrap again with foil and freeze for up to 6 months. Unwrap and thaw at room temperature for about 2 hours, then ice before serving.

Wine making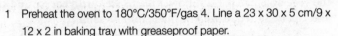

Carrot wine is surprisingly good, so if you have a glut of carrots give it a go. You could combine it with apples or parsnips.

o For 2 kg/4½ lb carrots use 4.5 litres/1 gallon water, 1.1 kg/ 2½ lb sugar, 450 g/1 lb chopped raisins and grated zest and juice of 2 lemons and 2 oranges.

o Boil the carrots in the water first until they are soft, then strain the liquid on to the sugar, raisins, oranges and lemons in the fermentation bin. Leave to cool and continue from Step 2 on page 63.

C CAULIFLOWER ▣ ✳ ▤

An attractive vegetable with a cream-coloured heart or curd, the cauliflower is a member of the cabbage family with varieties that can grow all the year round, cropping between June and November. Their leaves protect the curds from too much sun or frost, keeping them a glowing white. However, cauliflowers with purple, green or orange curds are readily available, making them a tasty novelty in the vegetable patch. Cauliflower teams well with cheese and spices and can be stored, frozen or made into piccalilli.

Varieties

o **All the Year Round:** As its name suggests, it can be sown in autumn or spring and is one of the easiest to grow producing large white heads suitable for storing.

o **Andes F1:** Lovely large blue-green leaves that protect the creamy white curd. A nutty hint to the flavour. Sow December to July to harvest May to October.

o **Gitano F1 Hybrid:** A nutritious Romanesco cauliflower with a pointed green curd. Harvest mid-September to November.

o **Graffiti F1:** A purple colour that intensifies with light so no need to protect the curd. Sow March to June, harvest June to October.

o **Lateman:** Ideal for growing closer together to produce small curds. Sow March to April to harvest July to October.

Cooking and preparing

Cauliflower must be harvested when the curds are tight and firm and early in the morning when the dew is still fresh on them. In winter, wait until the frost has melted. This attractive vegetable is glorified in Mediterranean and Eastern cuisine, used with herbs and spices to add texture to dishes and absorb flavours, especially in curries. Traditionally in the UK it is cooked with cheese sauce – adding sliced tomatoes on top before grating the cheese over, or stirring in cooked crispy bacon, or combining it with broccoli, makes it a delicious meal on its own which can be frozen. Stir-frying barely cooked cauliflower in butter and breadcrumbs gives it the Polish touch. Fresh young raw cauliflower has a sweet and nutty taste, perfect to include in a plate of crudités served with garlic mayonnaise.

C

o **To prepare:** Soak in salted water to remove any creatures, rinse and either cook whole cutting a deep cross at the base of the stalk, or snap off the florets to cook more quickly.

o **To cook:** To boil whole, lower into a pan containing about 4 cm/1½ in of boiling salted water, so that the curd steams. Cover and boil for 20 minutes until the curd is just tender. Florets on their own will take about 7 minutes in boiling salted water.

Storing

Cauliflower will store in the fridge for up to 2 weeks wrapped in clingfilm. Alternatively it will store for 3–4 weeks, dug out of the ground by its roots and hung upside down in a cool but airy dark place. Keep away from light to maintain its whiteness.

Freezing

o **To prepare:** After preparing as above, break the cauliflower up into even-sized florets.

o **To freeze:** Blanch for 3 minutes adding 15 ml/1 tbsp lemon juice to each 600 ml/1 pint water. Drain, dry and tip into freezer bags.

o **To thaw and use:** Cook from frozen in boiling, salted water for 5–7 minutes until tender.

o **Storage time:** 12 months.

Preserving

Traditionally, cauliflower is the main ingredients of the vibrantly-coloured mustard pickle, piccalilli:

Piccalilli

Some small green tomatoes, a little celery or a small cucumber (unpeeled) can also be added, if you like.

Makes about 4.5 kg/10 lb

3 large firm cauliflowers
900 g/2 lb pickling onions
450 g/1 lb courgettes
450 g/1 lb French beans
450 g/1 lb salt
4.5 litres/1 gallon water
225 g/8 oz sugar
15 ml/1 tbsp mustard powder
10 ml/2 tsp ground ginger
1.7 litres/3 pints vinegar
45 ml/3 tbsp cornflour
30 ml/2 tbsp turmeric

1 Prepare the cauliflowers as above and divide into small florets. Peel the onions and cut in half if they are too big. Slice the courgettes into medium-sized pieces. Thickly slice the beans.

2 Bring the salt and water to the boil in a pan and simmer until the salt has dissolved, then leave to go cold.

3 Put the prepared vegetables in a large bowl, pour over the salted water and leave to stand for 24 hours.

4 Drain and rinse the vegetables well.

5 In a large preserving pan, mix the sugar, mustard and ginger with all but 150 ml/¼ pint of the vinegar and bring to the boil. Add the vegetables and return to the boil and simmer for 20 minutes.

6 In a bowl, blend the cornflour and turmeric with the remaining vinegar until smooth and stir into the vegetables. Cook for another 3–4 minutes.

7 Remove from the heat and ladle the piccalilli into clean, warmed jars. Seal and label. Store in a cool, dark place.

CELERIAC ▣ ❋   **C**

An unusual-looking round, knobbly root, celeriac is related to celery and tastes very similar to it; the creamy white flesh has a nutty sweetness and subtle aniseed flavour. It is easier to grow than celery, but needs plenty of moisture in the soil to help the roots to swell. Celeriac doesn't suffer much from pests and diseases and stores well. It can also be frozen.

Variety

o **Monarch:** Good quality, creamy roots that are fairly smooth.

Cooking and preparing

Celeriac can be cooked in the same way as any root vegetable – mashed, chipped or roasted. It is particularly delicious mashed with butter and lots of freshly ground black pepper and eaten with roast meats or game. It is also very good raw, grated and mixed with mayonnaise in a dish called *remoulade* traditionally eaten as part of a plate of salads as a starter in France and Belgium.

o **To prepare:** Expect to lose about a quarter of the weight of each root in peeling off the tough knobbly skin. It is a very hard vegetable, so you need a very sharp knife – cut from the outside in, chopping it into manageable chunks, then peel each chunk rather than the whole root at once. Immediately place the peeled chunks into cold water with 15 ml/ 1 tbsp of lemon juice added to it, to prevent the flesh from discolouring, until ready to cook.

o **To cook:** Boil in a pan of salted water for 15–20 minutes until tender.

Storing ▣

Celeriac will keep in a cool, dry place for several weeks and for even longer if layered in sand or similar. Brush off the soil, trim off any green leaves and make sure the roots are dry before storing.

C Freezing ❄

Celeriac is not generally frozen as it stores so well, but it can be frozen grated or cut into thick slices or large dice. Prepare as above.

- o **To freeze:** Blanch in boiling water with 15 ml/1 tbsp of lemon juice for 3 minutes for slices or chunks, 1 minute for grated. Dry thoroughly and pack in convenient quantities in freezer bags.
- o **To thaw and use:** Cook the slices or chunks from frozen in boiling, salted water for 5 minutes until tender, 3 minutes for grated. Add from frozen to soups, stews and casseroles. Thawed grated celeriac can be used in salads.
- o **Storage time:** 12 months.

CELERY ▣ ❄ ◊

Celery is usually sold with pale stems meaning that they have been blanched – earthed up to keep out the light – while growing. Today there are many self-blanching varieties, which makes growing them simpler. However, if you grow it in the old-fashioned way in a trench, covered with insulation for protection, celery can be lifted right up to Christmas and beyond if mild. Celery's aniseed flavour is used to boost soups and stews and the crisp, crunchy vegetable is delicious raw. It can also be frozen to be cooked later.

Varieties
- o **Latham Self Blanching/Galaxy:** Good quality tender stems. Not prone to bolting.
- o **Victoria F1:** Crunchy, tasty, stringless stems with good bolt resistance. Self-blanching.
- o **Solid White:** A hardy trenching variety producing thick crisp stalks from November to February.

Cooking and preparing
Celery is popular with slimmers because it has virtually no calories but has plenty of fibre to help with digestion. It is delicious raw with dips or chopped up to give crunch to a rice salad or a bean salad and it cooks well as a base in stews and casseroles. It is also a very tasty vegetable cooked in its own right, braised in stock in the oven or on the hob in a covered sauté pan, and as soup.

o **To prepare:** Wash and trim off the root and the leaves; keep the latter to use as a herb. Pull off the stalks. If stringy, pull the strings off from the top along the stalk with a sharp knife. Slice or chop. To use the hearts whole, slice off the root end and pull away the thicker stems, leaving the heart intact.

o **To cook:** Slice and add to soups, stews and casseroles, sautéing with the onion at the beginning of cooking, or braise slowly in stock for about 30 minutes until tender.

Storing

Celery will keep for 2–3 weeks in the salad drawer of the fridge after harvesting, longer if stood upright in a vase of water. If you grow the trenching variety, they will keep in the ground with some insulation to protect against very heavy frosts.

Freezing

The flavour of celery intensifies after freezing, so be sparing when adding it to dishes. Select celery with firm heads.

o **To prepare:** As above, cut the stalks into even-sized pieces.

o **To freeze:** Blanch for 3 minutes. Pack well dried in freezer bags.

o **To thaw and use:** Cook from frozen. Once defrosted, celery goes soft so cannot be used raw.

o **Storage time:** 9 months.

Juicing

Celery juice is a nutritious and flavoursome component of any vegetable combination. The greener the stalks the more nutritious celery is. It is believed to have a calming effect on the nervous system and combined with apple may help relieve insomnia.

C CHERRIES ❄ ◐ ▢ ▤

Cherries have been grown in Britain since Roman times. A cherry tree is a wonderful thing to grow in the garden covered in blossom in the spring and shiny red or yellow fruit, dark or pale, in the summer. However, a late frost can destroy a potential crop. There are two types: sweet dessert varieties; and the sour cooking cherries that are the best for freezing and bottling, especially in brandy.

Varieties
Of the sweet varieties, the red cherries freeze the best.

o **Crown Morello:** A sour, self-fertile variety with large red fruits that preserve and freeze well. Doesn't mind the cold and will grow on a north-facing wall.
o **Kentish Red:** Self-fertile cooker that ripens in late July.
o **Summer Sun:** A sweet, succulent and plump red cherry ready for picking in late July.
o **Regina:** A heavy cropping, dark red, very sweet, juicy cherry resistant to splitting.

Cooking and preparing
Harvest cherries when they are glossy and soft to the touch although you may have some competition from the birds. If you can, drape a net over the tree before the fruit ripens. Cook sour cherries in pies, jams, syrups, liqueurs and sauces to accompany poultry and game; sweet, or dessert cherries, make good tarts.

o **To prepare:** Wash the cherries and pull them from their stalks; pit them before cooking, if you prefer.
o **To cook:** Simmer sour cherries with plenty of sugar to make sauces and pie fillings.

Freezing ❄
Frozen cherries keep their flavour well. Freeze sour varieties with sugar or in syrup.

o **To prepare:** As above and remove the pits if you prefer. If you leave them in they will impart a slight almond flavour.

o **To freeze:** Open freeze sweet red cherries on trays and return
   to the freezer in freezer bags. Open freeze sour cherries on
   trays mixed with 100 g/4 oz caster sugar for every 450 g/
   1 lb fruit. Pack cherries with sugar syrup to cover in a rigid
   container and freeze.
o **To thaw and use:** Cook from frozen.
o **Storage time:** 12 months.

Cherry Lemon Cake

Serves 6

200 g/7 oz butter
200 g/7 oz caster sugar
3 eggs, separated
Zest of 1 large lemon
450 g/1 lb plain flour
10 ml/2 tsp baking powder
75 ml/5 tbsp yoghurt
10 ml/2 tsp kirsch
350 g/12 oz fresh dark cherries
For the icing:
175 g/6 oz icing sugar
Juice of 1 large lemon

1 Preheat the oven to 180°C/350°F/gas 4.
2 Cream the butter and sugar together in a large bowl until light and
   fluffy. Stir in the egg yolks with the lemon zest.
3 Sieve the flour and baking powder together and fold into the egg
   mixture. Stir in the yoghurt and kirsch little by little in turns.
4 Whisk two of the egg whites until they form stiff peaks and fold into
   the mixture.
5 Wash, dry and stone the cherries, keeping them whole if possible.
   Fold gently into the mixture.
6 Pour into a greased and floured 24 cm/9½ in ring mould and bake
   in the centre of the oven for 1–1¼ hours, until a skewer comes out
   clean and the cake is shrinking from the sides a little.
7 Remove from the oven and let it rest for 5–10 minutes, then tip out
   on to a cake rack to cool.

C

8   Sieve the icing sugar into a bowl and add the lemon juice a little at a time, stirring continuously until the icing reaches a dropping consistency. Place the cake on a plate (cardboard if you are going to freeze it) and pour the icing around the top so that it trickles down the sides. Before serving, pile up the centre with more fresh cherries from your tree.

To freeze: After icing, open freeze, then double wrap in clingfilm with an extra wrapping of silver foil. Return to the freezer for up to 4 months. To use, unwrap and thaw at room temperature.

Juicing 🝆

Cherries make a healthy, tasty juice, but as it is very strong it is best combined with equal parts water or apple juice.

Bottling ▣

Cherries bottle well in syrup but are best in brandy or rum. Use sour cherries and don't remove the stones.

Preserving ▨

As cherries are low in pectin, add lemon or apple juice to the mix when making jam. Sour cherries make a more flavourful jam using 1.25 kg/2½ lb sugar to 1.75 kg/4 lb cherries plus the juice of 2 lemons. When the jam is boiling the stones will rise to the surface where you can scoop them out, so you don't have to pit them.

CHICORY 🝆 ❄

There are three types of chicory: the Belgian Witloof chicory, known as endive in France and the US, which is grown for its white, green-tipped chicon, or compact shoots, that appear when forced; a green, leafy, sugarloaf variety similar to lettuce; and the red-leaved radicchio. The chicons grow in the dark throughout winter and can be eaten raw as a salad and cooked. Some leafy chicory is hardy and can be harvested throughout the winter. They all have a slight bitterness to their flavour, but the chicons become quite sweet when cooked.

Varieties

C

- **Rossa di Verona:** A hardy radicchio that grows a pretty ball head of red leaves. Sow June to October to harvest throughout the winter. The colour of the leaves improves in low temperatures.
- **Sugar Loaf (Pain de Sucre):** A non-forcing green leafy type that can be harvested until January.
- **Witloof:** The classic forcing chicory.

Cooking and preparing

The non-forcing types of chicory have colourful bitter leaves that are best used in a mixed salad, whereas the chicons lose their bitterness once cooked. You can braise them in stock either in the oven or in a covered sauté pan, adding a pinch of sugar and a squeeze of lemon juice; or you can wrap them, already cooked, in slices of ham, cover them with cheese sauce, top with breadcrumbs and bake the gratin until bubbling and crispy. The leaves can be pulled off raw chicons and used to hold cold, diced vegetables mixed with mayonnaise, or they can be sliced into thin rounds for use in salads.

- **To prepare:** Use a stainless steel knife to cut chicory as it browns easily, cut out the bitter core at the base, wash and shred the leaves to add to a salad. To use the chicons raw, slice off the root end and cut out the bitter core with a sharp, thin knife; rub with lemon juice to prevent browning.
- **To boil:** Prepare the chicons but leave whole and add to boiling water with 5 ml/1 tsp sugar and 15 ml/1 tbsp lemon juice. Simmer for 20–40 minutes or until tender.

Storing 🎴

Newly harvested chicory will keep in the salad compartment of the fridge for up to 2 weeks. To force Witloof chicory for a supply of chicons throughout the winter:

1. If you have grown the chicory from seed, cut the plants back to ground level in autumn and use the leaves.
2. Dig up the roots and replant in pots of John Innes No 2 compost or in boxes of moist sand and store in a dark, warm place. Cover them with a black bin liner if necessary.
3. Harvest the shoots as they appear above the soil or sand and they will regrow.

C Freezing ❄

The Witloof chicory can be frozen raw but only if it is going to be cooked after defrosting. Select firm, white heads with yellowish edges, discarding any that are going brown.

- o **To prepare:** Wash and cut out the bitter core as above. Leave whole.
- o **To freeze:** Blanch for 3 minutes with some lemon juice added to the water, dry well, pack in freezer bags and freeze. To freeze a chicory gratin, prepare the dish in a covered foil container and freeze it before baking.
- o **To thaw and use:** Cook chicory heads from frozen in boiling, lightly salted water for 6–8 minutes until tender. Drain and serve. For a gratin, remove the lid, dot with butter and bake from frozen in a preheated oven at 200°C/400°F/gas 6 for 45 minutes until golden and heated through. Alternatively thaw at room temperature for 3–4 hours, then bake in an oven preheated to 190°C/375F/gas 5 for 35–40 minutes.
- o **Storage times:** Raw 6 months; cooked 3 months.

COURGETTES & MARROWS 🗆 ❄ ▤ 🍷

Courgettes and marrows are considered to be the beginner's plant as they are so easy to grow, as long as you have a rich, moist soil and are prepared to water them well. Courgettes grow in compact bush varieties or trailing. They grow quickly and produce masses of fruit, and as a bonus the large orange flowers can also be eaten, stuffed or deep fried.

Courgettes should be harvested when they are about 10–15 cm/ 4–6 in long and marrows are tastiest picked when about 25 cm/10 in long just before they reach maturity and their skin hardens. Courgettes freeze well and can be used in pickles and chutneys and if you are unable to harvest them for a while and return to some marrow-sized veg, you can turn them into jam or wine.

Varieties

- o **Marrow Green Bush:** A good all-rounder producing a large crop of green courgette-size fruits early in the season and green stripy marrows if left to mature, which store well.

o   **Defender F1:** Produces excellent yields of courgettes if you keep cutting the fruits.

o   **Cavili F1:** Pale green courgettes with creamy, flavoursome flesh. It is parthenocarpic, which means it will produce fruit even if the flowers are not pollinated.

o   **Gold Rush:** A vigorous, yellow-skinned variety of courgette.

Cooking and preparing

Young courgettes can be eaten raw in salads or halved and steamed and served with fish, or sautéed for a few minutes with butter and garlic for a vegetable side dish. Be careful not to overcook as they will go mushy. Add courgettes to bakes, tarts and vegetable stews such as ratatouille, thread them on skewers with other vegetables to barbecue, or roast them in the oven with onions, peppers and tomatoes to serve with couscous. Courgettes make delicious soups and the Italians cut them into sticks and deep fry them in a light batter. The flowers can also be deep fried in a tempura batter or stuffed with a mince mixture or cheese.

Marrows have a more watery texture and when old can become hollow, seedy and bitter. To eat straight from harvesting, the skin should be tender enough to be easily pierced with a fingernail. The tastiest flesh is close to the skin, so the centre can be scooped out and the marrow stuffed with a mince or rice mixture and baked. Peeled, deseeded and chopped into large chunks, it can be boiled or steamed and served with a parsley or cheese sauce or in a curry.

o   **To prepare:** Rinse courgettes well, top and tail and cut in sticks, slices, chunks or halves. Peel marrow, remove the seeds and cut up into manageable pieces.

o   **To cook:** Prepare courgettes and steam for 3–4 minutes, or stir fry in olive oil for 2 minutes, tightly cover with a lid and cook on a very low heat for another 3 minutes, shaking regularly until just tender. Alternatively, fry sliced courgettes in butter until soft. Steam marrow for 20–30 minutes until tender or wrap in foil and bake in a moderate oven until tender.

Storing

Marrows can be stored until Christmas in a cool, dry place either hanging in a string bag or hammock or on a slatted shelf not touching each other. However, they must be harvested when they are mature, so leave them on the plant until the stalk has dried and

C gone brown and the skin has toughened. Cut the marrow from the plant leaving a piece of stalk, then leave to dry in the sun or in a warm shed or greenhouse for about a week until they sound hollow when tapped. Wipe off any earth. Store at about 10°C/50°F and protect from frost.

Courgettes will keep for a few days in the fridge, but they will lose their flavour by the day.

Freezing ❄

Marrows are too watery to freeze but you can freeze courgettes if they are picked young and firm with brightly coloured, shiny skins. Trim the ends, cut into thick slices or sticks and blanch for 1 minute. Dry well and freeze in freezer bags. Store for 12 months.

Preserving 📧

Add courgettes in thick slices to piccalilli (page 114). If they are ready to harvest before the other ingredients, brine them and pickle them in vinegar to keep until needed. Marrow partners well with ginger and spices to make a delicious jam and chutney, especially if you have a large marrow to contend with.

Marrow Jam

This can also be used as a tart filling. For a sweeter jam, add the juice and zest of 2 oranges.

Makes about 2.25 kg/5 lb

1.75 kg/4 lb marrow, peeled and diced

1.5 kg/3 lb sugar

Juice and finely grated zest of 2 lemons

100 g/4 oz crystallised ginger, finely chopped

1   Layer the marrow in a bowl with the sugar. Leave to stand in a cool place for 24 hours.

2   Tip into a preserving pan with the lemon juice and zest and the ginger.

3 Heat gently, stirring until the sugar has dissolved. Bring to the boil and continue to boil for about 45 minutes until the marrow is clear and tender.

4 Put into prepared warmed jars, seal and label.

Marrow Chutney

Makes aout 1 kg/2¼ lb

900 g/2 lb marrow, peeled and diced
10 ml/2 tsp salt
1.2 litres/2 pints vinegar
15 ml/1 tbsp mustard powder
10 ml/2 tsp turmeric
10 ml/2 tsp ground ginger
100 g/4 oz sugar
2 onions, finely chopped

1 Sprinkle the prepared marrow with the salt and leave to stand for 24 hours.

2 Drain off the liquid and rinse. Put the marrow in a preserving pan with the vinegar, mustard powder, turmeric, ground ginger, sugar and 2 onions, and bring to the boil. Simmer until thick, stirring frequently, and spoon into prepared warmed jars.

Wine making 🍷

Marrow, again partnered with ginger, makes a tasty wine. You don't peel or deseed the marrow, just chop it all up into small cubes (2 kg/4½ lb) and put it in a fermentation bucket with the grated zest of 2 oranges and lemons, 25 g/1 oz root ginger and a crushed Campden tablet, then pour 4.5 litres/1 gallon boiling water on to it all. When lukewarm add the juice of all the oranges and lemons plus the pectin-destroying enzyme and, before adding the yeast starter, continue from Step 2 on page 63.

C CUCUMBER 🔲

If you have a greenhouse you can grow cucumbers all summer long for salads and cold soups. The more you pick the more will grow, so don't be afraid to crop them young and small for pickling. Ridge cucumbers and some of the smaller, prickly varieties are for outdoor growing and also pickle well when young.

Varieties

o **Carmen F1:** A prolific, all-female variety that has good disease resistance.
o **Cucino:** An all-female cucumber that can be grown inside or out, producing small fruits ideal for pickling.
o **Swing:** A new variety that is good for indoors and out. The slightly prickly fruit grow to about 20 cm/8 in long.

Cooking and preparing

Not only can cucumber be sliced and diced and eaten raw in salads and dips – chopped up in yoghurt with garlic and mint or dill – it can also be cooked in a soup that is served chilled for a summer starter, or poached, or fried, or stuffed and baked. Alternatively, it can be made into cucumber sauce, which can be frozen for 3 months. Cucumbers are too watery to freeze raw, but you can take advantage of this by freezing peeled chunks and using them instead of ice cubes in a drink. Cucumbers that have just been picked will store in the fridge for up to 2 weeks.

o **To prepare:** Peel, if you wish, and slice thinly or dice for salads; if the cucumber is slightly bitter, scoop out the seeds as well. Peel, scoop out the seeds and cut in chunks or small dice for cooking.
o **To cook:** Fry diced cucumber in butter until just soft, or poach lightly for a few minutes until soft.

Preserving 🔲

Pick cucumbers when they are young and small or cut larger, unpeeled cucumbers into chunks or slices that will fit in a pickling jar. Layer the cut cucumber with salt in a bowl and leave for 24 hours. Soak whole cucumbers in salted water for 2 hours. Drain, rinse and dry as well as possible. Pack into prepared jars with bunches of washed and dried dill and coriander seeds and/or mustard seeds. Cucumbers can also be added to piccalilli (page 114).

CURRANTS, BLACK, RED & WHITE

The blackcurrant and its close cousins red and white currants require little attention from the gardener. The newer varieties are more resistant to pests and diseases, but are still no match for the birds who relish the dangling strings of juicy berries, so they are best grown under netting, if possible. Late frosts can be a problem.

Blackcurrants, well known for their high vitamin C levels and other health-giving properties, are much sweeter than redcurrants. These have more delicate skins, so can be eaten straight from the bush, whereas the latter are mainly used in cooked dishes. White currants also produce sweet berries in long, creamy coloured strings.

All the berries freeze well and with their high pectin levels produce superb jams and jellies. They can also make a very healthy and tasty juice and syrup, which can be used in yoghurts, ice creams and sorbets, as well as wine.

Varieties
o **Ben Hope:** Popular, high yielding, pest and disease-resistant blackcurrant, with a good flavour.
o **Ben Sarek:** Disease-resistant blackcurrant, late flowering and so not affected by frosts. Produces huge tasty berries on neat, compact bushes about 1 m/3 ft high.
o **Jonkheer Van Tets:** Hardy, heavy-yielding redcurrant; an award-winning variety that crops in early July.
o **White Versailles:** A vigorous, heavy-cropping white currant.

Cooking and preparing
Turn the currants into juice or add to other fruit juices; a main ingredient in summer pudding, they also make delicious tarts, cooked or uncooked, and go well with other fruit in pies and crumbles. With a surplus of blackcurrants, jam is a must, and the sweet acidity of redcurrants translates into a fine jelly to accompany lamb and game and to enrich sauces and gravies. As the skins can be tough, it is best to cook the fruit well before adding any sugar if possible.

C

o **To prepare:** Snip the strings of currants from the plant with scissors when they are ripe but still firm. Pull the currants off the stem with your fingers or more efficiently, run down the strings with a fork over a bowl, discarding any that are over ripe or blemished. Pull or snip away the coarse tip of the currant. Wash if necessary and dry thoroughly. Currants can be kept in the fridge for several days.

Freezing ❄

Currants freeze well, although open frozen redcurrants collapse a little on defrosting.

o **To freeze:** Open freeze the prepared currants on trays. Pour into freezer bags when frozen and return to freezer. Alternatively, mix 450 g/1 lb fruit with 100 g/4 oz caster sugar and freeze in a rigid container, allowing roughly 1 cm/½ in headspace.

o **To thaw and use:** Defrost overnight in the fridge or for 3 hours at room temperature. Use as fresh.

o **Storage time:** 12 months.

Juicing ◊

All the currants, especially blackcurrants, make a deliciously healthy juice, but blackcurrants are not as juicy as the red and white, so juice them together. Preparation is minimal as the pulp, stalks etc will be left behind in the juicer. Use well-ripened fruit but make sure you discard any mouldy ones. Store in the freezer.

Currants make a good cordial as well, which will keep for several months in the fridge after opening.

Bottling ▣

Black and white currants bottle well in syrup, but make sure they have been well prepared and dried (see above) and be careful not to damage them when packing into the bottles. Try bottling them in rum (pages 41–43) to serve with ice cream or on top of cheesecake.

Preserving ▨

Blackcurrant jam is an all-time favourite and redcurrant jelly is a must for the store cupboard. Both are simple to make, although if you are making jam there is the lengthy task of topping and tailing the currants.

o **Blackcurrant jam:** Blackcurrants must be cooked in water for 20–30 minutes until the skins are soft before adding the sugar, otherwise the berries will remain hard in the jam. To make about 1 kg/2¼ lb jam, simmer 900 g/2 lb blackcurrants in 600 ml/1 pint water, then add 1.25 kg/2½ lb sugar, bring slowly to the boil and follow the instructions on page 48 from Step 5.

o **Redcurrant jelly:** There is no need to top and tail the redcurrants as they will be left behind in the jelly bag. For about 450 g/1 lb jelly, just cover 750 g/1½ lb redcurrants with water and follow the instructions on page 48 from Step 2.

Wine making ⏳

Black, red and white currants all make a good wine. They all need 2.25 litres/4 pints water adding to the fruit to make the must.

1 Mash 1.5 kg/3 lb fruit in the fermentation bin first.
2 Stir in the water and add 1 Campden tablet.
3 Dissolve 1.1 kg/2½ lb sugar in 1.75 litres/3 pints water and pour over fruit.
4 When cool, add 5 ml/1 tsp each pectin-destroying enzyme and tannin plus the yeast starter after doing the acid test and hydrometer and temperature readings. Cover well and leave to ferment for 7 days. Continue from Step 8 on page 63.

F FENNEL ☷ ❋ ◐

Apart from the wild, bitter fennel there are two types: sweet fennel whose feathery fronds and seeds are used as herbs (page 143); and Florence fennel, which has a tasty aniseed-flavoured swollen leaf base, or head, and is grown as a vegetable. It is delicious cooked, and makes a very healthy juice raw. Fennel is quite difficult to grow, because if it becomes too dry it can bolt before developing its swollen base. However there are some bolt-resistant varieties. Earthing up the stems as they grow and swell keeps them pale.

Varieties

o **Amigo F1:** A variety with resistance to bolting so can be grown earlier in the year.
o **Mantovano:** A vigorous plant that can be sown in the spring to harvest in the summer. Not prone to bolting.
o **Montebianco:** Large heads produced on vigorous plants. Sow June to August to harvest September to November.
o **Victoria:** A new variety that can be sown from April to July for autumn cropping. Heavy cropping with large smooth white heads.

Cooking and preparing

The crisp head of Florence fennel is not unlike celery, except that it has a sweeter, more distinctive aniseed flavour. Raw and sliced thinly, it is a great crunchy addition to salads and its feathery foliage can be snipped into the salad bowl, too. It can be steamed or boiled, sautéed (after blanching) or casseroled: cut into wedges, placed in an ovenproof dish with chicken pieces on top and some stock and garlic, makes a speedy, tasty supper dish; add lemon juice, garlic and olive oil to some lightly cooked fennel that's still warm, to allow the flavours to infuse before cooling, for a delicious but simple starter. Fennel – cooked, raw or as a herb – complements fish well.

o **To prepare:** Trim off most of the stalks as they are tough and trim off the fronds to use separately. Slice off the root end, but not too much as you want to keep the head in tact. Slice down the centre lengthways and cut into wedges for cooking or slice thinly for salads.
o **To cook:** Boil in slightly salted water for about 15 minutes until just tender, drain upside down. Steam for about 20

F

minutes. Braise with stock in the oven for about 30 minutes. To sauté, blanch first for about 8 minutes.

Storing 🖻

You can harvest fennel until November, earthing up the plants to blanch the head giving extra insulation against any early frosts. It will keep in the fridge for up to 2 weeks.

Freezing ❄

Fennel isn't at its best after freezing, but if you have a surplus it is the best place for a supply to add to casseroles and other dishes.

o **To prepare:** As above, cutting the fennel into quarters and trimming off the fronds.

o **To freeze:** Blanch for 3 minutes, dry well and freeze in freezer bags. Place the fronds directly into a freezer bag – there is no need to blanch them.

o **To thaw and use:** Chop the fronds and use from frozen as a herb. Cook the heads from frozen in slightly salted boiling water for 6 minutes until tender, or put straight into casserole dishes to be cooked from frozen.

o **Storage time:** 6 months.

Juicing 💧

Fennel is packed with nutrients and is said to be good for calming the nerves and digestion. It lends its strong aniseed flavour to juice combinations such as beetroot, carrot and apple.

── Apple Aniseed ──────

To make 1 serving, cut up 1 small fennel head (approx 100 g/4 oz) with 3 sweet apples cut into narrow wedges and process in the juicer. Drink at once or freeze.

F FIGS ☼ ❄ ▭ ▨

Fig trees are surprisingly hardy but to get a good crop you have to be cruel to them, planting them in poor soil and starving them of food and water. Their roots need to be restricted so they take well to being planted in a pot, but they do need a warm, sheltered spot to thrive.

British figs do not produce seeds as they do in their native south-west Asia and eastern Mediterranean, because figs are pollinated by the tiny fig wasp that is indigenous to the more southern warmer climes. A glut of figs, if too many to eat fresh, will be used up in no time – bottled in syrup, frozen and in preserves.

Varieties

Not all varieties are suitable for the British climate, but you have more choice if you have a greenhouse.

o **Brown Turkey:** An old favourite that is reliable grown outdoors and readily available from garden centres.
o **White Marseilles:** Another reliable favourite suitable for outdoor growing.
o **Rouge de Bordeaux:** A pale green fig with a purple flush and sweet red flesh to be grown in a greenhouse.

Cooking and preparing

Fresh figs are best eaten just over ripe when fragrant, sweet and soft to touch. In the Mediterranean, they are often paired with sweet, dry-cured hams, making a delicious combination. They can also be halved or quartered into savoury or fruit salads, roasted whole in foil parcels, poached in a sticky red syrup, made into tarts and ice cream or added to cakes. They make an excellent jam and chutney.

o **To prepare:** It is important to handle figs with care as they bruise easily. The skin is edible but if you prefer to peel it, plunge it into boiling water for 30 seconds and it will come off easily.

Drying ☼

Figs dry to a distinctive, concentrated, moist, sweetness, but those grown in the UK tend to shrink much more than their seedy Mediterranean cousins, resulting in small specimens. Drying them in an airing cupboard will take about 1 week.

Freezing ❄

Fresh figs freeze well if you select ripe fruit with undamaged skin.

o **To freeze:** Wash them carefully and dry on kitchen paper and either open freeze the fruit or freeze them in a light syrup. Poach the figs in the syrup for 4–5 minutes until tender but still holding their shape; cool quickly.

o **To thaw and use:** Thaw at room temperature for 2–4 hours. Poached figs may be reheated in their syrup and served hot.

o **Storage time:** With both methods, the figs can be frozen for 12 months.

Bottling ▣

Figs in syrup is a traditional Italian dessert and is perfect with ice cream or crème fraîche. Choose firm, unblemished figs and bottle either with or without their skins and add a squeeze of lemon juice to the syrup.

Preserving ▨

Fresh and dried figs can be used to make jam and chutney.

o **For jam:** If using fresh figs, you need 2.75 kg/6 lb fruit for 1 litre/1¾ pints water and 1.6 kg/3½ lb sugar, plus the grated zest and juice of 2 lemons.

o **For chutney:** Mix fresh figs with other fruits such as plums and use half-and-half balsamic vinegar with malt vinegar for a richer flavour.

G GARLIC 🔲 🔳

Garlic not only enhances savoury dishes, but also has medicinal properties – antiseptic and antiviral. It is easy to grow, although it takes a while. Plant the cloves in the autumn and you'll have garlic bulbs the following summer. The best way to store garlic is to dry it and it looks very attractive hanging up in plaits in the kitchen once it is dried. You can also eat fresh garlic straight from the ground, which has a milder, sweeter flavour.

Varieties

o **Albigensian Wight:** A large garlic from the southern region of France.
o **Elephant Garlic:** A large, mild bulb great for eating fresh and roasting whole.
o **Purple Wight:** Produces large chunky cloves and is slightly sweet.
o **Solent Wight:** The bulbs of this variety can grow to 6 cm/ 2½ in across. Stores well.

Cooking and preparing

Garlic is very pungent when raw, but can be crushed in small amounts into yoghurt, cucumber and dill for a Greek-style dip, mashed into butter to be spread on French bread and heated through, added to vinaigrette, or stirred into mayonnaise to dip into with raw vegetables or crudités. Roasts are enhanced by slivers of garlic being pushed into the meat before cooking, as is almost every other kind of savoury dish.

G

A head of garlic is divided into several cloves, each covered by a papery skin. The finer the cloves are chopped, the stronger they will taste. Whole cloves in their skins can be tossed into a tray of roasting vegetables 15 minutes before the end of cooking time for a mild sweet flavour and complete heads can be wrapped in foil and baked in a hot oven for half an hour.

o **To prepare:** Pull the cloves away from the head, slice off the attached end and peel off the skins. For a strong flavour in cooked and uncooked dishes, crush the garlic straight into the dish. Alternatively, roughly chop peeled garlic cloves and place them in a mortar with 5 ml/1 tsp of coarse sea salt, leave for 2–3 minutes then crush with the pestle into a paste.

Storing

Harvest garlic when the green leaves have died down and are brown and dry. Lay the heads in the sun for a few hours to dry out enough to brush off the soil, then plait in ropes and hang in a cool, dark, airy place until needed. When they have dried out enough to be used, you can bring a rope to hang in the kitchen and pick off the garlic as you need it. Freezing garlic is not recommended, as it loses its flavour.

Preserving

Garlic can be pickled in the same way as onions (page 153), and eaten the same way – far away from everybody!

Gazpacho

A cold, garlicky soup that makes a refreshing dish on a warm day and is a great way to use up any excess cucumbers and tomatoes.

Serves 4

1 large cucumber, peeled and finely chopped
1 green pepper, deseeded and finely chopped
1 red onion, peeled and finely chopped
6 tomatoes, peeled and finely chopped
4 cloves garlic, crushed
Juice of half a lemon
60 ml/2fl oz olive oil
2.5ml/½ tsp chilli powder
2.5ml/½ tsp shredded fresh basil plus a few whole leaves to garnish
Salt
Pinch of ground cumin
600 ml/1 pint tomato juice or the equivalent in blended, peeled fresh tomatoes

1   Place the cucumber, pepper, onion and tomatoes in a large bowl. Add the garlic and the remaining ingredients and mix well. Season to taste.
2   Place in the fridge overnight before serving. Garnish with some fresh basil leaves.
3   If you prefer, the soup can be puréed and served with a little of the finely chopped vegetables added to each bowl.

To freeze: Pour the puréed soup into a rigid container and freeze for up to 3 months. To use, thaw overnight in the fridge and serve with a sprinkling of fresh chopped herbs and croutons.

GOOSEBERRIES ❄ ▢ ▧ ▾   **G**

With a reputation for being tart, gooseberries can become quite sweet if they've had enough sun shining on them to ripen fully. However, traditionally it is plenty of sugar that does the trick. There are several varieties: the dessert red ones are the sweetest and can be eaten raw at their ripest; the green varieties have a strong, sharp flavour that is drawn out during cooking and enhanced by elderflowers. Gooseberries make good chutney, jam and wine.

Varieties

o **Invicta:** A culinary, green fruit with good resistance to mildew, a common ailment.

o **Pax:** A heavy cropping red dessert variety with good disease resistance.

o **Rokula:** Another red heavy cropper with good disease resistance.

Cooking and preparing

Gooseberries should be plump, green and slightly under ripe for cooking. Poach in a small amount of water with plenty of sugar and use in crumbles and pies, or puréed in ice creams and the traditional fool. For a delicious pudding, scatter red dessert gooseberries and sugar over a pre-baked sweet pastry tart case and pour over a mixture of eggs, cream plus 10 ml/2 tsp elderflower cordial and more sugar, then bake in a hot oven for about 25 minutes until the filling has risen and is golden. Gooseberry sauce goes well with fish, especially mackerel.

o **To prepare:** Snip off the tops and tails of the fruit with scissors and wash.

Freezing ❄

Gooseberries can be open frozen whole, topped and tailed, when they are still firm. When they are softer and well ripened, lightly poach them in a syrup and freeze them, or purée them first.

Bottling ▢

If you are growing both red and green varieties, bottle them together for an attractive as well as flavoursome result. Slice a very small

G piece off at either end of each gooseberry to prevent the fruit from shrivelling and bottle in a light syrup with 15 ml/1 tbsp elderflower cordial added or with some washed and dried elderflowers packed in with the gooseberries.

Preserving 🗒

Gooseberries are high in pectin so set well in jam, and marry wonderfully with spices in chutney. Add a few fresh, washed nettles to the jam while it is cooking to enhance the green colour and, to give a perfumed flavour, put some elderflower heads in a muslin bag and add to the jam for the last 15 minutes of cooking. Remove both the nettles and elderflowers before spooning into jars.

o **To make jam:** Prepare 1.5 kg/3 lb gooseberries and simmer gently in 600 ml/1 pint water until the fruit is soft. Add 25 g/1 oz butter and stir in 1.5 kg/3 lb warmed sugar. Continue simmering until dissolved, then boil hard to setting point.

─── Spicy Gooseberry Chutney ───
Makes about 1.5 kg/3 lb

450 g/1 lb gooseberries, trimmed

350 g/12 oz seedless raisins or chopped dates

225 g/8 oz onion, finely chopped

50 g/2 oz soft brown sugar

600 ml/1 pint cider vinegar

10 ml/2 tsp mixed pickling spice in a muslin bag

Salt

1 Stir all the ingredients together in a preserving pan and add the spice bag. Leave covered overnight.

2 Remove the lid and heat gently, stirring until the sugar has dissolved.

3 Boil gently for a few minutes, then reduce to a simmer, stirring occasionally. Watch the pan carefully and after about 20 minutes, when the liquid has gone and the chutney is soft, remove from the heat. If the chutney has stuck to the pan, add a little more vinegar.

4 Allow to cool and remove the bag of spice before spooning the chutney into clean, sterilised jars. Leave for a few weeks to mature.

Wine making 🍷

Gooseberries are ideal for making sparkling wine for a refreshing summer's drink the following year. Remember to bottle the wine in the tougher champagne bottles with a wire cage (muselet) over the stopper to prevent any nasty explosions.

1 Put 2.75 kg/6 lb washed, prepared green gooseberries in a fermentation bucket and squeeze the juice out of them by hand.

2 Boil 4.5 litres/1 gallon water and pour it over the fruit, cover and leave to cool.

3 Add pectin-destroying enzyme and a Campden tablet, cover again and leave to soak for 3 days, stirring daily.

4 Strain off the liquid through a fine mesh bag into another bucket. Test the acidity and add 5 ml/1 tsp citric acid if necessary. Stir in a sachet of wine yeast, yeast nutrient and 5 ml/1 tsp of tannin, cover well and leave to ferment for 4 days.

5 Continue from Step 7 Sparkling Apple Wine (page 73).

GRAPES 🔲 ☀ ❄ ⬤ ▤ 🍷

With conservatories being so popular, built to capture every last ray of sun, it makes sense to grow a grapevine inside and watch the grapes develop and hang down from the branches. However, a grapevine will also thrive outside in a sunny, sheltered spot, preferably against a south-facing wall, and can be grown very successfully in a large tub, which can stand on a warm patio in the summer for ripening and be kept in the cold but out of the frost and wind in the winter, after being heftily pruned.

One successful vine should give you around 6.8 kg/15 lb fruit a year – plenty to make a few bottles of wine or juice, or to dry into raisins or sultanas to use in your Christmas cake.

G Varieties

There are many varieties of grape, those for growing outdoors or under cover; white or black ones; dessert, or those more suited to being turned into wine or juice. The following can also be grown in tubs:

o **Black Hamburg:** One of the best-known varieties for growing indoors. Large black grapes that dry well into raisins and make a good red wine and juice.

o **Boskoop Glory:** A delicious black, outdoor-growing dessert grape suitable for wine. Harvest September to October.

o **Siegerrebe:** An outdoor grape that produces a good dessert wine. Ripens to a golden tawny colour in September. Needs a warm, sheltered spot and can also be grown in a tub.

o **Thompson's Seedless:** A dessert white grape suitable for wine and drying into sultanas. Should be grown under cover.

Cooking and preparing

A very versatile fruit, grapes are best picked fresh and used in savoury salads, fruit salads, dessert toppings, with cold chicken and mayonnaise, but they can also be cooked with fish, roasted with poultry and game or baked in fruit crumbles. Sharp-tasting grapes can be cooked until caramelised and sweet. Traditionally, grapes pair well with cheese, embellishing the cheese platter. They can also be made into jelly.

o **To prepare:** Wash in cold water and gently shake to dry. To peel grapes, plunge in boiling water for 2 minutes and slip off the skins. To remove the seeds, scoop them out with a small pickling fork or halve and remove.

Storing 🔲

G

Freshly picked bunches of firm, seedless grapes can be stored with the cut end of the stalk in water for 6–8 weeks if kept in a cool place, or in the fridge unwashed in an airtight container or plastic bag; the more loosely packed bunches can be hung in a net in a cool, dark, airy place for several weeks, but they must be cut when they are completely dry to avoid rotting. Keep an eye on them though and remove any grapes that show signs of deterioration.

Drying ☼

White grapes are dried to make sultanas and black grapes dry into raisins. They dry very well but many of the varieties that are best grown in Britain have seeds in them, so you will have crunchy raisins or sultanas if deseeding them is too much of a task. Dry them on closely spaced wire trays so that they don't fall through as they shrink and be careful not to over dry them.

Freezing ❄

Grapes don't usually freeze well as they don't have much substance to them and their skins can become tough, so try a few first before putting your whole harvest in the freezer.

o **To prepare:** Select plump, sweet fruit, without any discolouring around the stalk end, wash and halve them and remove the pips but keep seedless varieties whole.

o **To freeze:** Pack the grapes in rigid containers in a light syrup and freeze.

o **To thaw and use:** Thaw in the fridge overnight or at room temperature for 2–3 hours. Use in cooked dishes or fruit salads.

o **Storage time:** 12 months.

G Juicing

Grape juice is easily digested and is full of health-giving properties; it also freezes well. Choose very ripe grapes but make sure you remove any mouldy ones and those that are discolouring around the stalk end. Grape juice made from unripe grapes and called verjuice can be used instead of lemon juice in cooking and for acidulating water to keep fruit or vegetables from going brown. Verjuice can contain a blend of unripe grape juice, crabapples and other unripe fruit.

Preserving

Grapes, red or white, set quite well in a jelly, but need added pectin in the form of apples and/or lemon juice if they are seedless. To make 600 ml/1 pint jelly you will need 1 kg/2¼ lb grapes. The result will probably be quite tart, depending on how sweet the grapes are. Use with roast meats and game.

Wine making

Wine made from grape juice needs little help, as you might expect. Use a wine or fruit press to extract the juice and stir in 750 g/1½ lb sugar to 4.5 litres/1 gallon juice. When it has dissolved add a Campden tablet and cover and leave for 24 hours. Add a sachet of wine yeast and pour into a fermentation jar. Continue from Step 7 on page 63.

HERBS ☀ ❄ ▦　　　**H**

Most herbs not only provide flavour and are medicinal but are also pretty plants, so they can be grown throughout the garden in the flower borders or in a patch not too far away from the kitchen, or in lots of pots against a sunny wall of the house. Some herbs such as sage, thyme and rosemary, do not die down in winter so can be cut when needed throughout the year.

Basil is a very tender plant and is best grown in pots; coriander is also a tender plant and tends to go to seed quite quickly, giving a short harvest for the leaves, but they are quickly followed by the flower seeds which have a spicy flavour of their own and can be dried and stored. Dill also has the added bonus of seeds along with flavoursome frond-like leaves that have a similar aniseed flavour to bronze fennel, a hardier plant that dies back in the autumn to reappear in the spring.

Marjoram and oregano are different types of the same plant and have very pretty flowers and are easy to grow, coming in many colours. They are hardy perennials that can be divided up in the autumn – they are essential to Mediterranean cooking. Mint is so easy to grow that it can be a pest, spreading underground and popping up everywhere that you don't want it in the spring. There are many varieties with many different flavours but it is best to keep this plant under control in a pot. Parsley, a biennial, will limp on through the winter and revive with the warmer days. There are two types of tarragon, French – the more aromatic, but not so hardy – and Russian that is more vigorous and survives winter well. Bay leaves grow as bushes, trees and hedges.

The best ways to store herbs that don't have a continuous crop is to dry them or freeze them. Either way you will have herbs for the taking whenever you need them.

Cooking and preparing

A bountiful supply of herbs in the garden will ensure that your dishes will never be lacking in flavour and individually they make wonderful sauces to accompany dishes, such as chive and cream sauce to go with baked trout, parsley sauce for ham, mint sauce with lamb, sage with pork, chicken and tarragon, etc. Bunches of herbs can be snipped into salads, including the flowers, and they can be used to flavour vinegar, as well as for tea infusions either picked straight from the garden or dried.

H o **To prepare:** Pick herbs for storing just before they flower and preferably in the morning. Wash them well and dry in a salad spinner or on kitchen paper; pull the leaves from their stalks and tear up into dishes, snip with scissors or chop with a special two-bladed herb chopper.

Bajan Seasoning

This is an excellent way of using herbs and is cooked with fish and in soups and stews throughout the Eastern Caribbean. Smear it inside whole fish before baking, stir a small spoonful into fish mixture when making fishcakes and add a spoonful into any other dish you feel might need some spicing up. It will keep in a jar in the fridge until it is gone.

Makes about 600 ml/1 pint

185 g/6 oz spring onions with green tops
1 head of garlic
450 g/1 lb onions
Large handful each of fresh thyme, marjoram, parsley, basil, coriander and tarragon
15 ml/1 tbsp medium curry powder
15 ml/1 tbsp fine sea salt
15 ml/1 tbsp celery salt
15 ml/1 tbsp paprika
10 ml/2 tsp freshly ground black pepper (or to taste)
2.5 ml/½ tsp ground cloves
1 Scotch Bonnet red pepper or 3 chilli peppers
5 ml/1 tsp ground cumin
5 ml/1 tsp ground turmeric

1 Roughly chop up the spring onion and peel and chop the onions and garlic.
2 Wash and dry the herbs well and pick off all the leaves from the stems of the thyme, marjoram, parsley and tarragon.
3 Place the onions, garlic and herbs in a food processor with the rest of the ingredients. Process for about 5 minutes.
4 Store in the fridge in a clean, sterilised jar (page 47).

Drying

There are several ways to dry herbs but the best is tying them up in bunches and hanging them upside down in a cool, airy space. The seeds, such as coriander and dill, can be caught in brown paper bags and dried in the oven. Once dried they can be stored individually in glass jars, or you can make some bouquets garnis by wrapping a mixture of a few teaspoonfuls of dried herbs in muslin, or similar, tying it up and adding it to stews, then removing it before serving. Tarragon loses a lot of its flavour when dried, so it is best to freeze it. You can dry chives on a tray, but they keep their colour better when they are frozen.

Freezing ❄

Fresh herbs can be frozen as soon as they are picked. They freeze well and are convenient to use in cooking, although they are not suitable to be used for garnish after thawing. Just wash and pack each herb into individual plastic bags and freeze. Once frozen you can crumble off what you need from the stalks straight into dishes. Freeze the pretty blue borage flowers in ice cube trays and plop them straight into drinks.

Storage time: 6 months.

Preserving ▦

Vinegar absorbs flavours well, so herb vinegars are a good way of using up herbs and adding extra flavour to salad dressings. Herbs will also flavour olive oil well but use healthy, dried herbs (with no moisture or moulds) rather than fresh to avoid the growth of dangerous bacteria; leave to infuse for a few weeks before using.

KALE ▣ ❄ ◗

Kale is a member of the cabbage family with curly dark green leaves, packed with iron giving a strong flavour. It is a hardy plant and keeps going through the winter, improving after the first frost. It can be used in the same way as cabbage (page 105).

K Varieties

Kales vary in type from varieties bearing bright green or purple curly leaves to those with long, slender, dark green leaves.

- o **Black Tuscany:** Also known as black cabbage, this is an attractive kale with long, slender, crinkly, deep green leaves and a wonderful flavour. Can also be harvested when young and small to add to a bowl of salad leaves.
- o **Redbor F1:** A beautiful variety with deeply crinkled purple-red leaves. Interplant it with another green variety for a stunning show.
- o **Starbor F1:** A crinkly green kale that will grow well in a cramped area.

Cooking and preparing

A vegetable to be enjoyed throughout the winter, kale is best when eaten quite young and after the frosts have begun. The old leaves can be quite bitter. Eat kale as a vegetable side dish, steamed or lightly boiled, or perk it up by stir-frying it in sesame oil and garlic, then adding 15–30 ml/1–2 tbsp of water, covering and cooking it for a few minutes until tender; serve with a splash of soy sauce and toasted sesame seeds.

- o **To prepare:** Discard dry or discoloured leaves. Fold the leaves in half along the stalk, lay flat on a chopping board and cut or tear the stalk away. Wash thoroughly and shred.

Storing

Kale can remain in the ground throughout winter. Otherwise it will keep in the fridge for about 3 days.

Freezing

Blanch washed, whole leaves for 1 minute and shred when cool before freezing in freezer bags. Cook from frozen in boiling, slightly salted water for 6–8 minutes. It will keep in the freezer for 6 months.

Juicing

Kale is very nutritious but it is too strong to drink as a juice on its own and is best diluted with apple juice.

KOHLRABI 🔳 ❄️

K

Kohlrabi does not look anything like a cabbage, but it is a member of the same family. Although it resembles a root vegetable and acts like one, it is the swollen base of the stem that is harvested and eaten, either boiled whole or chopped up, a little like turnip. The leaves and stems can be eaten, too. It is easy to grow in a light, sandy soil and matures in a couple of months.

Varieties
o **Lanro:** A reliable white (light green) variety with a delicious sweet flavour.
o **Purple Danube F1:** An attractive purple colour, the globes are sweet and can be eaten raw in salads.

Cooking and preparing
A mild-flavoured vegetable, kohlrabi is best harvested when it has grown to between golf ball and tennis ball size. It can be used like turnip (page 203) and cooked in stews, soups and stir-fries. When young and sweet, it can be grated into coleslaw. One tasty way to cook it is to peel and chop the swollen base and boil it with a chopped carrot in enough chicken or vegetable stock to cover for about 8 minutes until tender; stir in some chopped parsley, a squeeze of lemon juice, a knob of butter and 10 ml/2 tsp of honey, and turn up the heat, stirring, to reduce. Boil fresh, young leaves and stems until tender.

Storing 🔳
Once mature, kohlrabi will remain in good condition for a week in the fridge, or 2 weeks if wrapped in a moist towel. The longer they are kept the more woody they are likely to become.

Freezing ❄️
o **To prepare:** Cut off the stalks and leaves, peel and cut into even-sized chunks.
o **To freeze:** Blanch for 2 minutes and pack in convenient quantities in freezer bags.
o **Storage time:** 9–12 months.

L

LEEKS 📷 ❄

The leek is the perfect plant to grow for a constant supply of fresh veg throughout the winter. Once it matures it can stay in situ until needed and very little care has to be given to it. It is the easiest of the onion family to grow, tolerating a wide range of winter conditions and you can extend the harvest season by growing different varieties to mature at different times.

Leeks provide a delicious accompaniment to all types of meat and fish and also make tasty gratins, soups and tarts.

Varieties

With a succession of varieties you can be harvesting leeks from September until April. Earthing them up as they grow creates longer lengths of the sweet white stem.

- o **Bandit:** A new, excellent variety with good resistance to bolting and rust. Harvest right through to spring.
- o **Carlton:** An early variety with long white stems that can be cropped from September to November.
- o **Musselburgh:** A popular mid-season leek that produces thick, straight stems and crops through winter.
- o **Pancho:** An early maturing variety that can be left in the ground and will remain in good condition over the winter.
- o **Toledo:** A reliable, vigorous leek that can be lifted from November to late February.

Cooking and preparing

The tender, white base part of the leek has the better, sweeter flavour, but the more fibrous green tops are tasty, too, with a stronger onion flavour. They make a difference to slow-cooked dishes such as soups, stews and casseroles or can be used to make stock. Picked young, the tender white stems can be served cooked and cooled with vinaigrette, baked or grilled whole and served with cheese sauce, or sliced and sautéed then mixed with soured cream and toasted flaked almonds. Leeks go particularly well with smoked fish, gammon and parsley sauce and in savoury tarts.

- o **To prepare:** Remove any coarse or damaged outer leaves and trim off the ragged tops and the roots. Cut a long slit down from the top to halfway down the white without cutting the leek in half and open it out under cold running water, peeling the leaves back to wash any soil and grit away. Drain or shake the leeks upside down. Slice the green tops thinly but if the white stems are not too fat, leave them in larger sections or keeep them whole.

o **To cook:** Steam young leeks whole for 5–7 minutes; boil larger leeks in chunks in salted water for 8–10 minutes; blanch trimmed leeks for 3 minutes then braise in stock for 1 hour; stir-fry thinly sliced leeks for 8–10 minutes until soft or sauté for 3–4 minutes then cover tightly and continue to cook for another few minutes shaking the pan occasionally.

Storing 🍳

The beauty of leeks is that they can be kept in the ground until needed, but if the temperature threatens to plummet and freeze the ground, lift them and put them in a bucket with soil to cover or store layered in boxes of sand, or similar, in a shed. They will keep in the fridge for a week or more.

Freezing ❄

You can freeze your leeks if you are unable to keep them in the ground over winter or would like some ready prepared vegetables to add to a stew or to make a quick soup. Young leeks freeze best.

o **To prepare:** Trim and wash as above.
o **To freeze:** Blanch whole leeks for 4 minutes, chopped in chunks for 3 minutes and sliced for 2 minutes. Drain and dry well and when completely cold pack into freezer bags and freeze.
o **To thaw and use:** Cook from frozen, and if using whole leeks cover them with a sauce.
o **Storage time:** 6 months.

L ——Leek Gratin——————————————

Serves 4

4 medium leeks, washed and trimmed
25 g/1 oz butter
50 g/2 oz soft cheese
1 small egg, beaten
60 ml/4 tbsp plain yoghurt
60 ml/4 tbsp freshly grated Parmesan cheese
Salt and freshly ground black pepper
Pinch of ground coriander
45–60 ml/3–4 tbsp fresh breadcrumbs

1 Preheat oven to 160°C/325°F/gas 3 and butter a shallow ovenproof dish or covered foil dish if freezing.
2 Bring the leeks to the boil in a pan and simmer until tender. Strain and cut lengthways, then across into pieces. Arrange in the dish.
3 In a bowl, mix together the butter and soft cheese. Now add the egg, yoghurt, half of the Parmesan cheese and seasonings. Mix lightly and spoon over the leeks.
4 Combine the remaining Parmesan with the breadcrumbs and sprinkle over the top. Bake for 20–30 minutes until bubbling and brown.

To freeze: Bake until the gratin is just starting to brown, remove from the oven and leave to go cold. Cover and freeze for up to 3 months. To use, preheat the oven to 180°C/350°F/gas 4, remove the cover and bake from frozen for 30–40 minutes until heated through, bubbling and brown. Turn up the heat for the last 10 minutes to brown it.

ONIONS & SHALLOTS

The onion is a universal vegetable providing the basis for so many savoury dishes throughout the year. It is a wonderful vegetable in its own right as well, exuding so much sweetness that it caramelises with long, slow cooking. Shallots are from the same Allium family and are sweeter and milder than the onion. They develop a head of several bulbs, whereas onions normally form one. Onions and shallots store extremely well, but can be prepared and frozen for convenience.

Varieties

Planting onion sets close together will result in small bulbs perfect for pickling. Shallots store for longer than onions, for between 9 and 12 months.

o **Ailsa Craig:** A popular onion that produces large bulbs and stores well over winter.

o **Hi Keeper:** A good Japanese variety of onion to sow in late summer/early autumn for overwintering as it is very hardy but does not store well.

o **Red Baron:** Superb red onion with a lovely mild flavour that stores well.

o **Paris Silverskin:** A tiny pickling onion that is delicious raw in summer salads.

o **Red Sun:** Red-skinned shallots with white-fleshed bulbs, excellent for pickling and for lengthy storage.

o **Pikant:** A good early, high-yielding shallot with many strongly flavoured bulbs that also stores well.

⚫ Cooking and preparing

Onions and shallots can be used in a variety of ways either raw or cooked, on their own as a dish – stuffed with couscous or rice, cheese, ham and tomato and baked in the oven – or cooked long and slowly until the onion's natural sugars and starch are released and begin to caramelise before being made into French onion soup.

The milder red onions, thinly sliced, go well in salads and stronger flavoured onions can be doused in boiling water after being chopped to reduce their strength, if you want to use them raw. Soups, stews, casseroles are usually created on a base of sliced or chopped onions fried until just transparent, whatever the main ingredient. They are also delicious roasted whole in their papery skins, or fried and made into gravy.

o **To prepare an onion:** Peel the onion and, with a very sharp knife, slice horizontally down the middle while squeezing either end of the bulb to prevent it from falling open and releasing its pungent vapours into the air (caused by sulphur compounds in the onion's cells) and making your eyes water. Quickly turn the cut sides of the onion face down on to the chopping board and slice downwards, again squeezing either end of the half to stop the slices falling open. To chop, turn the sliced half around, still holding it together tightly and keeping it face down and slice across again.

o **To prepare a shallot:** As these are smaller and milder, slicing shallots is not such a tearful business. With a sharp knife, slice off the root end and peel off the skin, then squeezing either end, slice the shallot horizontally, give it a half turn and slice across lengthways for a finely chopped shallot.

Storing 🖼

When the onions and shallots start to mature, from July to September, the leaves begin to yellow and wither. After this has happened, lift the bulbs during a dry, sunny spell, shake off the soil and leave them to dry on the surface for a couple of days, or in a warm, dry place under cover if it is damp. The more they dry out, the better they keep. Hang them in a dry, cool, airy place in nets or plaited together by their leaves, but watch out for dampness, which will lead to rotting. Onions will store until the spring and shallots for 9–12 months.

Freezing ❄

Make sure that you double wrap and seal onions and shallots very well before freezing, as their odour will taint everything else in the freezer. Choose firm bulbs with stiff, papery skins. Any that appear damp or are rotting or sprouting are not worth freezing.

o **To prepare:** Peel, then leave small pickling onions and shallots whole, and slice or chop larger ones.

o **To freeze:** Blanch chopped bulbs for 1 minute, sliced for 2 minutes, whole for 3. Drain and dry well. Pack into freezer bags and place them in rigid containers to prevent cross-contamination.

o **To thaw and use:** Defrost whole bulbs at room temperature for about 2 hours and either thaw sliced or chopped onions or shallots or cook straight from frozen as required.

o **Storage time:** 6 months.

Preserving 🍱

Good, firm small-sized onions make wonderful pickles traditionally to be eaten with a doorstep of bread and strong Cheddar and washed down with a pint of real ale. Make your own spiced vinegar, as well (page 53).

Red Onion Marmalade

This delicious 'jam' will keep up to 3 months and is great with hard cheeses and cold meats, especially turkey on Boxing Day.

Makes 1 kg/2¼ lb

75 g/3 oz butter
10 ml/2 tsp olive oil
1 kg/2¼ lb red onions, peeled and thinly sliced
75 g/3 oz light brown sugar
15 ml/1 tbsp freshly chopped oregano
Salt and freshly ground black pepper
375 ml/13 fl oz red wine
150 ml/5 fl oz red wine vinegar
30 ml/2 tbsp balsamic vinegar

1 Melt the butter and oil in a large vinegar-proof, heavy-based pan (page 52) on a high heat, and stir in the sliced onions, sugar and oregano.
2 Reduce the heat a little and season with salt and pepper. Cook uncovered, stirring occasionally to prevent the onions from sticking and burning, until they begin to caramelise and are so soft they break when pushed against the side with a wooden spoon, after about 30 minutes.
3 Stir in the wine and vinegars and simmer uncovered over a high heat for another 15–20 minutes, stirring occasionally, until the onions are a deep brown and the liquid has reduced so that you can see the bottom of the pan when a spoon is drawn across it.
4 Allow to cool for 10 minutes and spoon into warmed, sterilised jars (page 47), seal and label.

PARSNIPS 🔲 ❄ 🌢 🍷

A grossly underestimated root vegetable, parsnips come into their own in the winter, growing sweeter and more flavoursome once the frosts start. They keep well in the ground until you need them, or in boxes if the ground is too hard, and they make delicious warming soups, purées and chips, perfect for suppers on cold, dark, winter evenings. Although they take quite a lot of space up in the vegetable plot from March, when they are sown, until November, when they mature, you can plant quicker maturing vegetables between the rows.

Varieties

o **Avon Resister:** Can be grown close together, about 7.5 cm/ 3 in apart. Has good canker resistance.

o **Dagger F1:** Smooth roots and good canker resistance – ideal for mini roots.

o **Gladiator F1:** A hybrid parsnip, which means it should produce more uniform plants and roots of excellent quality. Has a lovely smooth, white skin and a very sweet flavour.

o **Javelin F1:** Another variety that can be harvested for mini or full-grown roots.

o **White Gem:** Heavy yields and highly resistant to canker.

Cooking and preparing

Parsnips can be used like potatoes: roasted, boiled, chipped, puréed and fried. They can also be turned into crisps, along with beetroot, potatoes and carrots, or roasted in the oven on a tray that includes other root vegetables with onion and cloves of garlic and a scattering of rosemary. Parsnip soup is a very popular dish and partners well with apple. Parsnips can also enrich meat stews and casseroles, as well as being an important ingredient in vegetarian dishes. Like carrots, parsnips go well in bread and cakes.

o **To prepare:** With a sharp knife, cut off either end of the parsnip and check how woody the core is; if it seems dry and tough, halve the parsnip lengthways and remove the core. Thinly peel the roots or scrub them well.

o **To cook:** Cut parsnips into chunks and boil in lightly salted water for 10–15 minutes, a little less for slices, blend into purée with some butter, milk and a pinch of ground nutmeg. Steam parsnips for 15–20 minutes.

P ## Storing 🎴

Keep them in the ground all winter if it does not freeze too hard to dig them up. Otherwise lift them and store them layered in boxes of sand or similar.

Freezing ❄

You shouldn't need to freeze parsnips as they benefit more from being dry stored, but if you lack the space for storing, freezing is an option. Choose young firm roots.

o **To prepare:** Scrape or peel, then cut into slices or wedges.
o **To freeze:** Blanch for 2 minutes, drain and dry well. Freeze in freezer bags.
o **To thaw and use:** Cook from frozen in boiling, lightly salted water for 6–8 minutes. Alternatively, blanch from frozen for 2–3 minutes and roast in the oven in your usual way.
o **Storage time:** 12 months.

Juicing 💧

As parsnips have such a strong flavour, it is a good idea to blend the juice with apple or other vegetable juices.

Wine making 🍷

Parsnips have a very good reputation where wine making is concerned, producing a good flavoured result. However, they can be troublesome during the process, being prone to throwing a pectin or a starch haze, which appears in the fermented wine (see When the Wine Won't Clear, page 65). Parsnip wine should be left for at least a year before drinking.

—— Parsnip Wine————————————————

1   Scrub and dice 2 kg/4½ lb parsnips and simmer very gently (to prevent a starch haze forming later on) in 4.5 litres/1 gallon water until tender.

2   Strain the liquid on to 450 g/1 lb chopped raisins in a fermentation bin and cover.

3   When cool, add 5 ml/1 tsp each of pectin-destroying enzyme and tannin and continue with Step 3 on page 63.

PEACHES & NECTARINES ☼ ❄ ◖ ▢

Peaches and nectarines lose their flavour very quickly once picked, so they are at their perfect best eaten straight from the tree when they are warm, heavy and juicy. Peaches and nectarines are closely related but peaches are the more stalwart of the two, giving good yields on a sheltered sunny wall in a mild year but not much in a cold year. The smooth-skinned nectarines need to be grown in a warm conservatory, polytunnel or greenhouse to be sure of success. If you are lucky enough to have a glut, the best way to preserve them is to bottle them.

Varieties

The following are the most reliable and popular varieties of peach and nectarine to be grown in the British climate, but they do still rely on a mild spring and warm summer.

o   **Peregrine:** Grown in Britain for more than 100 years, this flavoursome, white-fleshed peach is a heavy cropper, especially on a warm, sheltered wall.

o   **Rochester:** A yellow-fleshed peach that flowers later than Peregrine, so there is less risk of frost damaging the blossom. Produces large, juicy fruit.

o   **Fantasia:** A red-skinned yellow-fleshed nectarine that can grow outside on a sheltered wall and crops in August.

o   **Lord Napier:** a white-fleshed, tasty nectarine that needs a well-protected spot.

P Cooking and preparing

If you harvest the fruit under-ripe, place them on a sunny windowsill to soften. Do not store them in the fridge as they will remain hard, bland and floury. Slice perfectly ripe fruit on top of tarts and pavlovas, mixed into fruit salads or added to green mixed salads; lightly poach peaches and nectarines in syrup and serve with vanilla ice cream and raspberry sauce for the classic Peach Melba dish, or bake in halves in the oven with a ground almond stuffing. They can also be served as an accompaniment to roast pork as a sauce or a stuffing.

o **To prepare:** Be very careful when picking peaches and nectarines not to bruise them. Cup the fruit in the palm of your hand, lift and twist gently; if it doesn't come away easily, leave it for another couple of days. To peel, plunge into boiling water for 1 minute, then into cold and with a sharp knife, slip off the skin, halve and remove the stone and cut into quarters if necessary.

Drying ☼

Peaches and nectarines can be dried but they are not as good as dried apricots as they do lose some flavour. Halve and stone the fruit first.

Freezing ❄

Peaches and nectarines can be frozen, but again they lose flavour in the process. Remove their skins and stones and freeze them in halves or slices in syrup. Add lemon juice to the syrup for the white-fleshed fruit to prevent discoloration. Freeze in rigid containers for up to 12 months. Thaw in the fridge and use as required.

Juicing ◌

The juice of these fruits goes off very rapidly so they are best puréed and frozen immediately, then thawed in the fridge and diluted with water or other fruit juices as required.

Bottling ▣

Select perfect fruit, just ripe, and remove their skins and stones and cut them in half before bottling in syrup. Peaches and nectarines can also be bottled in brandy, rum or vodka (page 44) for a delicious dessert with ice cream or a digestif after a sumptuous meal.

PEARS ▣ ☀ ❄ ◗ ▭ ▨ ♀

Pear trees have long lives, cropping well for 50 years or more. They are from the same family as the apple (*Malus*) but they have a milder more subtle flavour and a very short time when they are perfect to eat: hard and flavourless one day, perfect the next then cotton wool the day after, so it is important to keep an eye on your pears as harvest time approaches. Pears can be stored if picked at the right time and are delicious bottled; perry pears can be made into perry, similar to cider.

Varieties

The main groups are culinary (cooking) and dessert (eating), the latter being a better choice for growing in the garden. Perry pears are not generally suitable for smaller gardens and they can also take 10 years to fruit.

- o **Beth:** A vigorous early dessert pear cropping in September. Suitable for small gardens. Not self-fertile.
- o **Concorde:** A self-fertile late dessert variety cropping in October/November, also suitable for cooking.
- o **Conference:** The most popular dessert variety, which also cooks well; self-fertile.
- o **Williams' Bon Chretien:** An easy-to-grow hardy dessert variety with a great flavour and very juicy. Crops in September and good for bottling, but does not store well. Both Beth and Conference will pollinate it.
- o **Warden:** a large culinary pear excellent for bottling; harvest in October and store until February.

P Cooking and preparing

When soft, sweet and juicy, pears are the perfect accompaniment for any sharp blue cheese. This, combined with bitter leaves and walnuts, can create a delicious salad. Late-season dessert pears such as Concorde cook well and are especially good for desserts. Divine when drizzled with dark chocolate, in tarts or simmered in sweetened red wine. There are numerous delicious recipes that use pears and they can often be a substitute for apples in pies, flans or for baking or making purées and chutney, although they have a milder flavour.

o **To harvest:** Pears should be harvested when nearly ripe, unless you are going to eat them straight away. Most tend to turn from a very acid green to a slightly lighter hue when ready for picking and some acquire a rosy blush, but colour is not always a good guide. To pick, simply lift and gently twist; if they come off without resistance they are ready.

o **To prepare:** Before cooking, cut the pear in half lengthways with a sharp knife and cut again into quarters; cut out the core and peel, slicing off the stalk.

Storing 🖼

Harvest the pears when they are not quite ripe, keeping on their stalks, as being kept cool and dark will prevent them from ripening any further. Choose blemish-free fruit, taking great care not to bruise any and lay them out, well cushioned with paper, on trays. Bring them into the kitchen when you are ready for them and the ripening process will restart with the warmth.

Drying ☼

Do not dry pears in rings like apples but pick them when they are perfectly ripe, peel and core them dipping them straight into acidulated water (page 23) to prevent them from browning. Choose your method of drying, perhaps a faster method than air drying as the pear halves may be quite thick.

Freezing ❄

P

Pears are only worth freezing if they are poached first.

o **To freeze:** Prepare them as above and poach gently in a light syrup, replacing half the water with red or white wine or cider if you prefer – for 5 minutes for quarters, 10 minutes for halves or 15–30 minutes for whole ones depending on their ripeness. Turn the whole ones over halfway through poaching. Pack into rigid containers leaving some headspace but making sure the syrup covers the fruit.

o **To thaw and use:** Thaw in the fridge overnight or at room temperature for 3–4 hours. If serving hot, make sure the fruit is heated right through.

o **Storage time:** 8–12 months.

—— Pears in Burgundy ——————

Serves 6

300 ml/½ pint water
300 ml/½ pint red Burgundy
225 g/8 oz caster sugar
Strip of lemon or orange zest
5 cm/2 in stick of cinnamon
6 ripe pears
Toasted, flaked almonds to garnish

1 Place all the ingredients except for the pears in a large pan or casserole, large enough to fit the pears tightly standing up, and gently heat, stirring, until the sugar has completely dissolved. Boil briskly for 3 minutes without stirring. Remove from the heat.

2 Peel the pears as thinly as possible, leaving them whole with the stalk on.

3 Stand the pears up in the pan, cover, bring the syrup back to the boil and simmer very gently for 20–30 minutes, or until just tender.

4 Remove the pears and place them in a serving dish.

5 Meanwhile remove the zest and cinnamon stick from the syrup, turn up the heat and boil the syrup hard, uncovered, until reduced by about half.

P

6   Pour the syrup over the pears and scatter over the flaked almonds before serving, warm or chilled, with crème fraîche or double cream.

To freeze: Cool quickly and pack the pears and syrup in a rigid container. Freeze for up to 3 months. To use, thaw overnight in the fridge or at room temperature, uncovered for 4–5 hours. Serve garnished with the flaked almonds.

Juicing ⬡

Pears can be difficult to put through a juicer as they can turn into ooze, so make sure they are firm, ripe and sweet and juice the chopped up pieces alternately with apple pieces, starting and ending with the apple. While peeling and chopping, dip the fruit into acidulated water to prevent discoloration. Drink or freeze the juice immediately for maximum benefit.

—— Apple and Ginger Pear Juice ——

To make a serving, said to be good for the digestion, juice 1 eating apple with 1 dessert pear, then add 2.5 cm/1 in piece of ginger, sliced, followed by another apple. To make it worth freezing, multiply the ingredients and freeze in useable quantities.

Bottling ▣

Pears can be bottled very successfully. Use firm fruit that are only just ripe, peel and halve them scooping out the core. Dip them in lemon juice. Either poach them in syrup before bottling or pack the bottles with the prepared pears, being careful not to bruise them, and fill up with sugar syrup and other spices or flavourings.

Preserving ▤

Pears make a great pickle and chutney. Pickle prepared pears by lightly cooking them in a boiled solution of white vinegar, sugar and spices, such as cinnamon, allspice, ginger and cloves. Drain and pack the pears into sterilised jars. Continue boiling the vinegar until it thickens and pour over the fruit in the jars, covering well.

───**Peggy's Pear Chutney**────────────── **P**

The pears can be a little under ripe but not too much, and they must be firm.

Makes about 3.5 kg/8 lb

2 kg/4½ lb pears
450 g/1 lb onions, finely chopped
10 ml/2 tsp salt
750 g/1¾ lb sugar
15 ml/1 tbsp ground ginger
450 g/1 lb dates, chopped
225 g/8 oz sultanas
15 ml/1 tbsp mustard powder
1.2 litres/2 pints vinegar

1 Peel, core and chop the pears.
2 Put the pears and onions into a pan with the remaining ingredients.
3 Bring to the boil and simmer until thick and brown, stirring frequently. How long this takes will depend on the ripeness of the pears.
4 Spoon into prepared jars, seal and label.

Perry making 🕓

If you have a treeful of perry pears use them to make perry in the same way as apples are made into cider. It is not worth making perry with cooking or dessert pears as it will not taste pleasant.

If you have a William's Bon Chretien tree, you can grow a pear inside a bottle and preserve it in brandy or vodka, making a delicious and impressive digestif.

1 Single out a young promising looking pearlet and fix a bottle to the tree so that the pear can be placed inside it and continue to grow. Attach the bottle to the tree upside down so that moisture doesn't get in and plug the opening with a piece of gauze to prevent insects from crawling up but still allow a free flow of air.

2 When the pear is ripe – and make sure it doesn't over ripen – carefully wash the inside of the bottle and pear with boiled cooled water and tip it upside down to drain.

3 Fill it nearly to the top with brandy or vodka, or similar, push a cork halfway into it and store in a cool, dark place for 3 months.

P

4   Top up the bottle with more alcohol or sugar syrup (page 44), gently agitating the bottle to mix, push the cork back in and leave for at least another 3 months before drinking.

PEAS ☼ ❄

The best way to enjoy peas is straight from the plot. In fact, you would have to have grown a very large crop for any fresh young peas to even make it as far as the kitchen, they are so tasty just picked and popped straight into your mouth. The thin, flat, mangetout varieties and fatter sugar snap peas are even better value, as you can eat the pod as well. Sow seeds in succession until the end of May for a continuous crop throughout the summer and freeze the surplus as soon after harvesting as possible to safeguard their high vitamin C content and sweetness. If you are unable to pick them all, just leave them on the plants to dry, although they won't be as tasty and nutritious as frozen peas, but at least they won't be wasted.

Varieties

o   **Alderman:** A late maincrop variety that grows to about 1.5 m (5 ft) high and has long pods packed with peas.

o   **Delikett:** A heavy cropping, stringless sugar snap pea with a very sweet flavour.

o   **Early Onward:** A high-yielding maincrop variety that benefits from early sowings.

o   **Feltham First:** A popular dwarf variety that is particularly good for sowing very early or late in the season. Doesn't need much supporting.

o   **Greensage:** Bred from the popular Greenshaft variety, but said to be even sweeter.

o   **Pea Oregon:** An easy-to-grow heavy cropping mangetout that grows to around 100 cm/3 ft high with 11 cm (4½ in) long pods.

Cooking and preparing

Peas are so delicious picked young and eaten straight from the pod, or whole if they are sugar snaps or mangetouts. Lightly boil podded peas and serve them with mint and butter as a side dish as they come or crush or purée them. Add them to risottos, rice salads, tuna and pasta bake or anything that could do with an extra tang of sweetness. Make pea and mint soup using the pods as well by simmering them in stock until soft and blending with cream or a béchamel sauce to thicken if necessary, or use frozen peas – this can be frozen for up to 3 months; thaw at room temperature and gently heat through. Mangetout and sugar snap peas can be steamed or added to stir-fries. Add dried peas to stews and casseroles.

o **To prepare:** The beauty of mangetouts and sugar snap peas is that after washing and snipping off the stalks, no more preparation is needed, except that larger mangetouts may need stringing. Shelling peas can be laborious but it is worthwhile and the pods are good for the compost heap.

o **To cook:** Simmer for a few minutes in unsalted, boiling water – salt will toughen their skins – or sauté in butter until soft. Steam or stir-fry mangetouts and sugar snap peas.

Drying ☼

Maincrop peas that have been left too long on the plant will become hard and tough, leave them until their pods have gone yellow and the peas have become hard inside. Pick them and pod them discarding any with holes in them made by the maggots of the pea moth. Make sure that the peas are completely dry before storing them in clean dry jars. Before using, cover them with boiling water and leave to soak for a couple of hours, before boiling, or add without soaking to stews, soups or casseroles.

Freezing ❄

For sugar snaps, mangetouts and maincrop varieties, select tender young peas in smooth, shiny, crisp pods, avoid any that look grey or shrivelled. Make sure that the sugar snaps are nicely plump. Prepare as above and shell the maincrop varieties.

o **To freeze:** Blanch each type for 1 minute and open freeze. Pour into freezer bags and return to the freezer.

o **To thaw and use:** Cook each type from frozen in lightly salted boiling water for 3–5 minutes until just tender. Mangetouts and sugar snaps can be added frozen to stir-fries.

o **Storage time:** 12 months.

P **PEPPERS** ☀ ❄ ▨

Peppers, sweet or chilli, need a long, very warm and sunny growing season to succeed, so in the UK it is best to grow them in a greenhouse or polytunnel for a reliable crop, large enough to store. Also known as capsicum, sweet peppers are commonly either green, red, orange or yellow – both the green and the orange mature to red; the green ones are more acidic and not so sweet. They are delicious sliced and grilled or stuffed, or used in salads. There is a vast array of chilli pepper varieties, ranging from the mild to fiercely hot – they make very attractive plants and along with sweet peppers look lovely drying in long ropes or ristras (page 25).

Varieties

- o **Tasty Grill Red F1:** A sweet pepper that produces long red fruits up to 25 cm/10 in in length. There is also a yellow form, Tasty Grill Yellow F1.
- o **Big Banana F1:** Really sweet fruits of up to 25 cm/10 in long that mature to a bright red.
- o **Gypsy F1:** Masses of sweet peppers up to 10 cm (4 in) long and 8 cm (3 in) wide that turn orange to red. Crops earlier than some varieties.
- o **Jalapeno Summer Heat F1:** A chilli pepper with long narrow fruits that can be picked when green or left to mature to red.
- o **Numex Twilight:** An attractive small chilli pepper to grow in a pot on the patio. The masses of tiny fruits ripen from purple to yellow, orange and red and all these colours can be on the plant at once.
- o **Thai Dragon F1:** A truly hot variety that is a prolific cropper of 9 cm/3½ in red fruits.

Cooking and preparing

Eaten raw, sweet peppers add juicy, crunchy texture and colour to green salads, sliced into rings. They are wonderful roasted or barbecued, charring the skin black, which you can then peel off. These soft, sweet, roasted peppers can be sliced and tossed through pastas, salads and made into sauces. Their bulbous shape and size makes them ideal for stuffing and baking using any combination of rice, cheeses and fresh herbs, or mince. Use chilli peppers to put some heat in sauces, curries, stews and casseroles, particularly in Mexican dishes.

P

o **To prepare sweet peppers:** Halve lengthways and cut away
   the stalk, seeds and white membrane. The skin is tough and
   does not break down during cooking so you can peel it off
   by rubbing the pepper with oil and placing it under a hot
   grill until blistered, then seal in a plastic bag for a minute or
   two and the skin will come off easily.

o **To prepare chilli peppers:** Be very careful when preparing
   chilli peppers as if you get the juice on your hands and then
   touch your eyes or any other sensitive parts of your body, it
   will be very painful. Either use rubber gloves or wash your
   hands thoroughly after preparing them. Slice the chillies
   down the centre and remove the seeds, turn the halves cut
   side down and slice thinly. Scrub the chopping board well
   after use.

Drying ☼

Both sweet peppers and chilli peppers dry very well and look
attractive hanging up in ristras. To use, slice or crumble into soups,
stews and casseroles.

Freezing ❄

Sweet peppers freeze well without blanching but they will not keep
for long, whereas chilli peppers will keep for a year. Select firm,
shiny peppers, wash and freeze whole in freezer bags or halve them,
remove the stalks and seeds, slice and freeze.

o **To prepare:** Cut off the tops of sweet peppers and remove
   the seeds or leave them whole. Prepare chillies as above or
   freeze them whole.

o **To freeze:** Blanch whole peppers for 3 minutes, sliced or
   diced peppers for 1 minute. If you would like to freeze
   stuffed peppers, fill them with a stuffing of your choice
   after blanching. Seal the peppers in freezer bags and freeze.
   Open freeze prepared chillies and double wrap them before
   returning to the freezer.

o **To thaw and use:** Cook from frozen. Halve, deseed and slice
   whole chillies while still frozen. When using ready-prepared
   chillies, 2.5 ml/½ tsp is equivalent to 1 medium chilli.
   Use sliced or diced sweet peppers from frozen. While still
   frozen, deseed whole peppers, slice and cook immediately,

P

alternatively you can stuff them at this stage and allow an extra 4–5 minutes cooking time. Leave frozen stuffed peppers to thaw at room temperature for 2–3 hours, then cook in your usual way.

o **Storage times:** Unblanched sweet peppers 2 months; blanched 12 months; stuffed 3 months; chilli peppers 12 months.

Preserving 🗒

Pickle sliced sweet peppers in a hot vinegar and sugar solution and chillies in a cold pickling vinegar or add them to a medley of pickled vegetables.

PLUMS ☀ ❄ ◐ ▭ 🗒 🍷

Having a plum tree in your garden has to be good value. More often than not it will be laden with plums towards the end of the summer and if you can't eat them all straight from the tree, there are so many great ways of preserving them – they dry and freeze well, are delicious bottled and make wonderful jams, pickles, chutneys, sauce and liqueurs. Damsons and gages from the same *prunus* family can be similarly preserved. Plums need very little care, just a well-sheltered sunny spot and moisture-retaining soil.

Varieties

There are three types of plum, the sour, firm culinary plum, the sweet, juicy dessert plum and the dual-purpose plum; and there are many varieties in a medley of colours – yellow, green, red, dark blue, purple and black – varying in texture, tartness and sweetness.

o **Victoria:** Popular, heavy-cropping dessert plum with large reddish fruit that freeze and bottle well.

o **River's Early Prolific:** A sweet, juicy dual-purpose plum with purple skin and yellow flesh. An early cropper.

o **Oullins Golden Gage:** A large, sweet golden-fruited member of the gage family that grows well in the UK and is delicious straight from the tree or cooked.

o **Marjorie's Seedling:** A large blue-black dual-purpose plum. The longer it is left on the tree the sweeter it becomes. Excellent for cooking and preserving.

P

- o **Merryweather Damson:** This grows well in exposed wet areas where other plums might fail. It has blue-black fruit with a true damson flavour.
- o **Giant Prune:** Also known as Burbank, this sweet, dark red plum dries into delicious prunes.

Cooking and preparing

Small, firm or under ripe plums are best for cooking. Reduce them into a sharp, sticky sauce to accompany duck, venison or game, or bake them in crumbles, tarts, pies and cakes.

- o **To prepare:** Firm plums are easier to prepare, just halve them and scoop out the stone with a sharp knife. The skin can be quite tart; to remove it, plunge the fruit into boiling water and then into ice-cold water and slip off the skins.

Drying ☀

All plums will dry into prunes, some better than others. Choose perfect specimens and wash, halve and stone them. Dry them slowly on wire racks to preserve as much colour as possible.

Freezing ❄

Plums freeze very well; either open freeze or dry-pack them with sugar or freeze them in a heavy syrup, or stew them first and purée if you wish. You may want to remove the skins as they can be tough and tart when thawed. It is also best to remove the stones as they can taint the overall flavour.

- o **To freeze:** Wash, halve and stone the plums, lay out on baking trays and open freeze. Return to the freezer in freezer bags. Alternatively, pack in layers in a rigid container sprinkling about 175 g/6 oz caster sugar to 450 g/1 lb fruit between them, seal and freeze. Freeze stewed or puréed plums in freezer bags or rigid containers.
- o **To thaw and use:** Put open frozen plums straight into crumbles, pies, sauces, etc. Thaw dry-packed plums or those in syrup or stewed at room temperature for 2–3 hours and use as required.
- o **Storage time:** 12 months.

P Juicing ◻️

Wash, halve and stone the fruit before juicing. If the result is too thick, dilute with water, apple or grape juice. If your plums do not produce enough juice, chop and heat them slowly in a covered pan with a little water. Once the fruit has broken down, strain, sweeten and bottle or freeze.

Alternatively, turn your plums or damsons into a cordial and pour over ice cream or dilute with sparkling water to drink.

Bottling ◻️

Choose perfect, blemish-free purple or red plums or damsons that are a little under ripe, wash them and prick them with a needle through to the stone. Bottle them in syrup or in a mixture of three-quarters syrup and one-quarter alcohol, such as brandy or kirsch, and use the slow water bath method; leave to mature for 2 months before using. Alternatively, preserve whole pricked plums or damsons in just rum or brandy (page 44).

Preserving ◻️

All types of plums make wonderful preserves and plum sauce. Damsons make excellent butters and cheeses to serve with meats. When making plum jam, crack open some of the stones and add the almond-flavoured kernels to the pan to flavour during cooking.

—— Green Plum Chutney ——

Makes about 3 kg/7 lb

2 kg/4½ lb stoned unripe plums
2 large onions
450 g/1 lb raisins
450 g/1 lb sugar
450 g/1 lb cooking apples, chopped
15 ml/1 tbsp salt
5 ml/1 tsp pepper
10 ml/2 tsp ground ginger
15 ml/1 tbsp mustard seed
25 g/1 oz pickling spice
600 ml/1 pint vinegar

1 Use green unripe plums and stone them before weighing. Cut up the plums and chop the onions finely.

2 Place in a pan and mix with the raisins, sugar, chopped apples, salt, pepper, ginger and mustard seed.

3 Tie the pickling spice in a muslin bag and boil in the vinegar for 6 minutes, then remove.

4 Pour the spiced vinegar over the other ingredients and simmer gently until brown and thick. Spoon into warmed jars, seal and label. Leave to mature for 3 months.

Plum Sauce

Delicious with duck and Chinese dishes.

Makes about 1.75 kg/4 lb

1.5 kg/3 lb ripe plums
750 g/1½ lb golden syrup
900 ml/1½ pints vinegar
15 ml/1 tbsp ground ginger
15 ml/1 tbsp salt
2.5 ml/½ tsp freshly ground black pepper
15 ml/1 tbsp ground cloves

1 Put all the ingredients together in a pan and bring to the boil. Simmer for 1 hour.

2 Put through a sieve. Reheat, pour into bottles or preserving jars, seal and sterilise using one of the bottling methods on pages 41–43.

P —— **Pickled Damsons** ——————————————

Makes 2.75 kg/6 lb

2.75 kg/6 lb damsons
900 ml/1½ pints vinegar
1.75 kg/4 lb sugar
10 ml/2 tsp pickling spice
6 cloves

1   Wipe the damsons, prick them with a needle and put into a bowl.
2   Boil together the vinegar, sugar and pickling spice, tied in a muslin bag with the cloves, for 5 minutes, then pour over the fruit.
3   Leave to stand for 24 hours, then drain off the juice and bring back to the boil.
4   Pour back over the fruit and leave to stand for another 24 hours.
5   Take out the spice bag and pour both the fruit and the juice into the pan. Bring to the boil slowly being careful to prevent the fruit from breaking up.
6   Lift out the fruit with a slotted spoon and carefully put into warmed, sterilised jars.
7   Bring the liquid back to the boil, pour over the damsons and seal. Leave to mature for 2–3 months.

Wine making ⧗

Both plums and damsons make a superb flavoursome wine. However, it is important to add the pectin-destroying enzyme, pectinase, in Step 1 to avoid a haze forming with either fruit. Also add 225 g/8 oz chopped raisins to 2 kg/4½ lb fruit at the same time to give more body.

POTATOES 🗖 ❄

P

If you have enough space to grow them, you will never have to buy another potato again and although they are cheap enough to buy, nothing beats the taste of freshly dug home-grown potatoes, early or maincrop. You can store them all through the winter until you need them, giving a permanent supply of potatoes that can be mashed, chipped, baked and made into gratins. Early or new potatoes are their most tasty eaten on the day they are dug and are not worth storing as they tend to lose their flavour with each passing day, but they can be frozen.

Varieties

There are so many varieties to choose from that can be planted throughout the growing season. They are divided into first earlies, the first new potatoes that can be enjoyed in June; second earlies, in July; maincrop, in August; and late maincrop, in September.

o **Arran Pilot:** A very popular first-early potato with a fantastic new potato flavour.

o **Epicure:** A good white first-early potato with creamy flesh. Best one to recover if touched by frost.

o **British Queen:** More than 100 years old, this second early variety produces good harvests of floury potatoes that have a delicious flavour.

o **Estima:** A heavy yielding second early. Drought resistant and one of the earliest varieties to produce good baking potatoes.

o **Desiree:** Probably the most popular red maincrop variety with waxy, pale yellow flesh. Great all rounder and heavy cropper. Makes great chips.

o **Golden Wonder:** A maincrop potato that keeps well and makes good chips.

o **Sante:** A good maincrop potato for pest and disease resistance. A dry, pale yellow flesh that boils and bakes well.

o **Belle de Fontenay:** A maincrop salad potato with a wonderful flavour.

o **Cara:** A good drought and disease-resistant late maincrop variety that makes a great baking potato.

o **Pink Fir Apple:** A red knobbly-shaped, late maincrop salad potato renowned for its delicious flavour. The shape

P

makes them trickier to peel but the flavour outweighs this inconvenience.

o **Sarpo Mira:** A new late maincrop variety that has proved to have really good blight resistance. Not attractive to slugs.

Cooking and preparing

First early new potatoes are best eaten straight from the ground, scrubbed and washed and either boiled or steamed with mint. You can leave the skins on but some can give a sour after taste, so scrub or scrape off the skins for optimum flavour. Older potatoes can be divided into two types – floury and waxy. Floury potatoes are usually late season and break up easily when boiled, this is due to the sugars converting into starch as the growing season lengthens – perfect for mashing, frying and roasting. Tempt your children with healthy chips – unpeeled potatoes cut into segments and roasted in a little olive oil. Waxy potatoes have a higher sugar content and firm texture that prevents them from absorbing too much water during cooking so that they retain their shape.

o **To prepare:** Resist peeling potatoes when possible as the nutrients are concentrated just under the skin. However, it is best to peel them for boiling if they have any green bits, which are poisonous, or other blemishes, which should be cut out with the sharp end of the potato peeler. If they are blemish free you can boil them in their skins and then peel them once cooked. Peel them for chips and roast potatoes. Scrub the skins well for jacket potatoes and roasted oven chips and cut out any blemishes.

o **To boil:** Place the prepared potatoes chopped into evenly sized chunks – larger pieces for floury potatoes – in a pan full of cold, salted water and bring to the boil. Simmer for 15–20 minutes depending on the size of the pieces until the sharp point of a knife can go in without resistance. Fast boiling will cause them to break up.

Storing

Dig up the potatoes rubbing off as much soil as possible, then leave them outside to dry for an hour or two. Select only blemish-free potatoes and when they are completely dry place them gently in heavy duty paper bags or, ideally, hessian sacks and store them in a cool, dry, dark place to prevent them from going green. Towards the end of winter, if they start to shrivel and sprout, increase the chill factor in the storage place but don't put them in the fridge.

Freezing ❄

o **To prepare:** Select freshly dug, small new (early) potatoes and scrape off the skins first. Boil until nearly cooked and cool quickly. Deep fry chips in batches for 2 minutes until soft but not brown; drain well.

o **To freeze:** Pack new potatoes in rigid containers with melted butter and mint. Freeze when cold. Open freeze chips and pack in freezer bags.

o **To thaw and use:** Thaw the new potatoes at room temperature for 2–3 hours, then heat gently. Deep fry chips from frozen until crisp and golden brown. Drain on kitchen paper.

o **Storage time:** Partly cooked new potatoes, 3 months; chips, 4 months.

Q QUINCES ▣ ❄ ▨ ♟

To the Ancient Greeks and Romans the quince was referred to as the Golden Apple. In fact it is part of the same family of apples and pears but they are too sour to eat fresh from the tree. When ripe, the golden-coloured quinces have a strong, sweet smell and a honey-like flavour and pink flesh when cooked, but they are still very hard, although they may soften after a frost. The quince is high in pectin so makes wonderful jellies, jams and cheeses and just a couple of slices can make a difference when combined with other fruit, such as apples, in a pie.

Varieties

There are two types of quince, the ornamental quince, or Japonica (*Chaenomeles*). This has an abundance of lovely red, pink or white flowers in spring that develop into small, edible, quince fruits in the autumn. They grow best against a wall or fence. The true quince (*Cydonia Oblonga*) bears larger brighter yellow fruit that have 10 times more flavour than the Japonica and just one quince can fill a room with its fragrance.

o **Knap Hill Scarlet:** An ornamental Japonica with large orangey-red flowers in the spring that produces yellow fragrant fruit.

o **Meech's Prolific:** Pale pink blossom and pear-shaped golden yellow fruit in October to November. Self-pollinating.

o **Quince Vranja:** A vigorous tree that can grow to 3.6 m/ 12 ft. White blossom develops into large, round golden yellow fruit.

Cooking and preparing

Leave quinces on the tree until they are completely ripe and are about to drop. As the fruits are so hard, chop them as best you can and make the traditional Spanish Membrillo – a quince cheese similar to fruit butter (page 49) – by boiling them with a small amount of water, then either sieving or blending the soft pulp to a purée; bring the purée to the boil again with its equal weight of granulated sugar, stirring until thick and coming away from pan; spread out on rice paper-lined Swiss roll tins and leave, lightly covered, in a warm place to dry; cut into pieces, sprinkle with caster sugar and store in an airtight container.

Quinces soften quite quickly when boiled and their skins just seem to melt away; only a couple of thin slices are needed to add extra flavour to cooked apples. Removing the cores is quite hard work, so try to avoid it.

Storing

Quinces can be stored in a cool, dark place for 2 months or so, but must be kept in a different area from other produce as their strong aroma will taint. Choose perfect specimens and keep an eye out for any bad spots.

Freezing

Freeze the purée in rigid containers or cook the washed and chopped up quince (cored if preferred) in a heavy syrup, with the pared zest of a small lemon, for 20 minutes until just tender, then freeze in rigid containers for up to 12 months. Cook from frozen.

Preserving

Quince in Portuguese is *marmelo* and marmalade was originally quince jam in Britain in medieval times. If you don't want to peel and core 2 kg/4½ lb quinces to make jam, you can boil them whole, covered, for 1½–2 hours first with some of the pre-weighed sugar before lifting them out with a slotted spoon, cooling them enough to handle and removing the core and chopping them up; then return the quinces to the liquid with the rest of the sugar and, once it has dissolved, boil until setting point is reached.

Q — **Quince Jelly** —————————————————————

To make jelly, you just chop up the whole fruit.

Makes about 1.5 kg/3 lb

1.75 kg/4 lb quinces
3 litres/5¼ pints water
10 ml/2 tsp citric acid
Sugar

1 Chop the quinces into fairly small pieces and put into a pan with 2.25 litres/4 pints of the water and the citric acid.
2 Cover and simmer for about 1 hour until the fruit is tender. Strain through a jelly bag for 15 minutes.
3 Remove the pulp from the bag and cook with the remaining water for 30 minutes. Strain through the jelly bag again and combine both lots of juice.
4 Add 450 g/1 lb sugar for each 600 ml/1 pint juice and stir over a low heat until the sugar has dissolved.
5 Boil hard to setting point, pour into warmed jars and seal.

Wine making 🖵

Due to its strong aromatic flavour, the quince makes a good wine and is an excellent way of preserving a tree full of fruit. Wash the fruit well before chopping up and discard any rotten bits. Use chopped raisins and a pectin-destroying enzyme.

RASPBERRIES ❄ ◐ ▢ ▨ R

A summer favourite, raspberries are a delight to have growing in the garden – and with a cropping season that can span the summer into autumn, the more you grow the better. (The birds love them, too, so a fruit cage or netting is a good idea.) Any left over from fresh pickings can be so easily frozen without detriment and used as fresh in cakes, tarts or pavlovas or transformed into delicious raspberry jam, so different from that you can buy in the shops. The loganberry is a hybrid of the raspberry and blackberry and can be used in the same way (page 93).

Varieties

Raspberries are divided into two groups: those that fruit in the summer; and those that fruit from August to October. By choosing plants from each group, it is possible to be picking over a number of months. For example, you could choose:

o **Glen Moy:** One of the earliest varieties for July.
o **Glen Ample:** Follows the above in July to August.
o **Octavia:** A great variety for mid-July to late August.
o **Autumn Bliss:** A good reliable autumn fruiting raspberry from August to mid-October.
o **Allgold:** Sweet and yellow-fruited. Crops late August to mid-October.

Cooking and preparing

Raspberries are traditionally lauded for the wonderful jam they make. They are also delicious made into tasty tarts, crumbles with or without apple (good if you have plenty of frozen raspberries) and summer puddings, ice cream, sorbet, cheesecakes, muffins, mousses as well as with cream in a Victoria sponge or on top of a feathery, crispy pavlova. Their intense sweet flavour also lends itself well to game and rich meats.

As the raspberries ripen, the fruit separate from their core leaving an inner cavity so preparation is easy. Simply pick and rinse just before use, if you feel it is necessary. Cluster berries such as these have thin papery skins that make them fragile and perishable. Handle carefully, and try not to keep them piled on top of each other in a container while waiting to be used.

R —— **Raspberry Tart**——————————————

Serves 8

For the pastry:

200 g/7 oz plain flour

25 g/1 oz polenta

15 ml/1 tbsp caster sugar

150 g/5 oz butter

1 egg, beaten, plus extra for brushing

For the filling:

175 g/6 oz mascarpone

60–90 ml/4–6 tbsp crème fraîche

30 ml/2 tbsp caster sugar

15 ml/1 tbsp freshly grated orange zest

15 ml/1 tbsp freshly squeezed orange juice

450 g/1 lb fresh or frozen raspberries, thawed

1   Grease a 23 cm/9 in loose-bottomed flan tin and line the base with baking paper.

2   For the pastry, mix the flour, polenta, sugar and butter in a food processor until it resembles fine breadcrumbs, then add the egg to make a dough.

3   Tip the dough on to a floured surface and lightly knead into a flattened ball. Wrap in clingfilm and chill for 30 minutes.

4   Meanwhile, preheat the oven to 180°C/350°F/gas 4. Roll out the pastry to line the flan tin. Prick the base well and line with baking paper and baking beans. Bake blind for 10 minutes.

5   Remove the blind and brush all over the base and sides with beaten egg. Return to the oven for another 10–15 minutes until golden.

6   Leave to cool in the tin, then remove and place on a serving plate.

7   For the filling, blend the mascarpone with 60 ml/4 tbsp crème fraîche, sugar, orange zest and juice. Add more crème fraîche if necessary. Fill the tart with the mixture and smooth over the surface.

8   Spread the raspberries over the top.

To freeze: Freeze the pastry at Step 6. When completely cold, wrap carefully in foil and clingfilm and freeze up to 3 months. To use, unwrap and thaw in the fridge overnight. Preheat oven to 180°C/350°F/gas 4 and crisp for 10–15 minutes.

R

Freezing ❄

Select ripe fruit that haven't gone mushy. If over ripe they won't freeze well.

o **To freeze:** Spread raspberries over a baking sheet and open freeze. Pour into freezer bags and return to freezer. Alternatively, dry freeze layering the fruit in rigid containers with sugar. For a pipless purée, blend the fruit in a blender or food processor, pass the purée through a sieve and sweeten to taste before freezing in a rigid container.

o **To thaw and use:** Leave the open-frozen and dry-frozen raspberries to thaw at room temperature for about 2 hours and use as fresh, or cook from frozen. Leave purée to thaw for 1–2 hours or heat in a pan from frozen or thaw in a microwave.

o **Storage time:** 12 months.

Juicing 〇

Raspberry juice can be a bit tasteless so it is best mixed with apple juice to enhance the flavour. Raspberry cordial is excellent diluted with sparkling water or drizzled over vanilla ice cream.

Bottling ▣

Raspberries make a wonderful liqueur. Bottle them in white rum or brandy, then leave them to macerate until Christmas.

Preserving ▤

Raspberries are high in pectin and so are perfect for jam making and you don't have to boil it for long either, keeping the fruit as whole as possible. After opening, you will need to keep the jam in the fridge. Making jelly instead dispenses with the pips.

R ── **Fruity Raspberry Jam** ──────────────

This jam retains the flavour of the raspberries beautifully and is very quick and easy to make.

Makes about 2.75 kg/6 lb

2 kg/4½ lb raspberries
2 kg/4½ lb sugar

1   Warm the sugar (page 47). Place the raspberries in a large pan and slowly heat so that the juices run. Bring to the boil.
2   Take off the heat immediately and stir in the warmed sugar until it has dissolved.
3   Return the pan to the heat and bring back up to the boil, but only boil for just 5 minutes.
4   Pour into warmed jars, seal and label.

── **Raspberry Vinegar** ──────────────────

Raspberry vinegar is an old-fashioned remedy that was once given to children when they had a cold. You can make this with other soft fruit, too, such as loganberries, blackberries or blackcurrants. Dilute with cold sparkling water for a summer drink, or boiling water for a nightcap.

1   Cover the fruit with white vinegar in a large glass or ceramic bowl and mash it each day for 4 or 5 days, stirring well.
2   Strain and measure the resulting juice and allow 450 g/1 lb sugar to each 600 ml/1 pint juice. Boil for 10 minutes and pour into clean sterilised bottles.

RHUBARB ❄ ▦   **R**

One plant of rhubarb can soon become two plants that can then become four plants… if you have the space. Although officially a vegetable, as the colourful stalk, or stick, is the edible part, rhubarb is easy to grow and if you feed it well and divide it every three years you will get good value. By forcing rhubarb – popping an upturned bucket, dustbin or rhubarb forcer over the plant in January – you could be harvesting the slender sticks from February, which freeze well.

Generally, a rhubarb plant should not be harvested for longer than four months to give it time to recover but with different varieties cropping at different times you could be harvesting until October. Rhubarb's acid, individual flavour makes tasty jam and chutneys and is delicious with custard.

Varieties
o **Timperley Early:** Produces pale pink, thin sticks from mid-February without need for forcing.
o **Stockbridge Arrow:** An early modern, virus-free heavy cropper that is able to produce up to 2.75 kg/6 lb red-stemmed rhubarb per plant in a season and, if forced, juicy tender sticks of up to 60 cm/2 ft long.
o **Stockbridge Harbinger:** A new, mid-season variety suitable for forcing with long, upright, tender, juicy sticks.
o **Victoria:** A popular, later cropping rhubarb with long, thick red sticks.

Cooking and preparing
Rhubarb needs care and attention when cooking to prevent it disintegrating into a mush. It also needs sugar to counteract its acidity and marries well with ginger or orange. Rhubarb releases a lot of juice when cooked so always pre-bake the pastry case when making a rhubarb tart and before adding the raw or slightly cooked rhubarb and custard filling, sprinkle 30–45 ml/2–3 tbsp of ground almonds over the base to soak up the juices. Rhubarb also makes wonderful pies and crumbles, and try layering some just cooked rhubarb between the slices of bread in bread and butter pudding with a sprinkling of grated orange zest.

o **To prepare:** Pull the sticks gently from the plant, rather than cutting them and chop off the leaves. These are poisonous

R

but are fine to throw on the compost heap. Trim off the bottom end, wash and chop into 5 cm/2 in chunks.

Stewed Rhubarb

1   Stew rhubarb carefully so that it doesn't lose its form. Dissolve 175 g/6 oz of caster sugar in 150 ml/¼ pint water and bring to the boil.

2   Add 700 g/1½ lb prepared rhubarb with the juice and grated zest of 1 orange, cover and, on a low heat, gently allow to return to the boil, watching it carefully.

3   Simmer for no longer than 5 minutes. Remove from the heat and leave to stand, tightly covered, for another 10 minutes.

4   Remove the lid and cool quickly by plunging the base of the pan into cold water. Keep in the fridge and serve with plenty of custard.

Rhubarb and Ginger Crumble

Serves 6

700 g/1½ lb rhubarb, trimmed
50 g/2 oz soft brown sugar
Juice and zest of 1 orange
For the topping:
100 g/4 oz gingernut biscuits, crushed
50 g/2 oz rolled oats
50 g/2 oz demerara sugar
100 g/4 oz butter, softened

1   Preheat the oven to 180°C/350°F/gas 4.

2   Slice the rhubarb into 5 cm/2 in chunks and place in a pie dish mixed with the soft brown sugar and the orange juice and zest.

3   Pulse the topping ingredients together in a food processor to form a rough crumble and spread over the top of the rhubarb.

4   Bake in the oven for 30–40 minutes until golden. Leave to cool a little so that the topping crisps up and serve warm.

To freeze: Remove from the oven 10 minutes before the end of cooking time. Allow to go cold and double wrap the dish in foil. Freeze for up to 3 months. To use, thaw at room temperature for 4 hours, preheat the oven to 180°C/350°F/gas 4 and cook for another 20 minutes until the crumble is golden and heated through.

Freezing ❄

Open freeze young firm sticks of rhubarb, chopped into chunks, or blanch them for 1 minute and freeze in a heavy syrup with the juice and grated zest of an orange added if liked. Alternatively, cook in the orange syrup (as above) and freeze when cold. Either way, it will keep for 12 months.

Use the open frozen rhubarb from frozen, but thaw the rhubarb in syrup at room temperature for about 2 hours and use as required.

Preserving 📧

Rhubarb has a poor pectin content so needs a high-pectin fruit to go with it for jam, such as blackcurrants or lemon juice. Generally, the rhubarb is left to macerate in a bowl with the sugar for 24 hours before cooking. Try a rhubarb and marrow jam, using 1.4 kg/3 lb marrow with 700 g/1½ lb rhubarb, the juice of 3 lemons, 100 g/4 oz finely chopped crystallised ginger and 1.8 kg/4 lb sugar. Macerate the rhubarb and marrow in the sugar first.

──── Rhubarb and Rose Petal Jam ────

Makes about 750 g/1½ lb

450 g/1 lb rhubarb
Juice of 1 lemon
450 g/1 lb sugar
100 g/4 oz dark red rose petals

1   Chop the rhubarb and put it into a non-metal pan or bowl with the lemon juice and sugar. Leave to stand overnight.
2   Cut the white or yellow tips off the rose petals, chop them and add them to the mixture. Bring to the boil and stir over a low heat until the sugar has dissolved.
3   Boil hard to setting point. Pour into warmed jars, seal and label.

Rhubarb also makes delicious chutney, especially flavoured with ginger and garlic.

—Muscovado Rhubarb Chutney—

Makes about 2.25 kg/5 lb

1 kg/2¼ lb rhubarb
450 g/1 lb sultanas
1 kg/2¼ lb Muscovado sugar
1 large onion, chopped
10 ml/2 tsp ground ginger
10 ml/2 tsp salt
2.5 ml/½ tsp freshly ground black pepper
2 lemons
600 ml/1 pint vinegar

1　Cut the rhubarb into small pieces and put into a pan with the sultanas, sugar and onion, ginger, salt and pepper.
2　Peel the lemons and remove the pips. Cut the lemon flesh into small pieces and add to the pan.
3　Pour in the vinegar and bring to the boil. Stir well and simmer until brown and thick, stirring frequently.
4　Spoon into warmed, sterilised jars, seal and label.

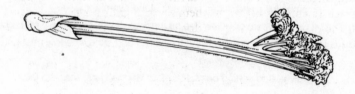

SALSIFY 🔲 ❄

Salsify is a long white root, rather like an elongated parsnip, which has an unusual, delicate taste that many say is akin to that of an oyster, hence its moniker vegetable oyster. The roots grow to 18–30 cm/9–12 in long and up to 5 cm/2 in diameter at the top. Being part of the daisy family, the plants produce attractive feathery leaves and purple daisy-like flowers 45 cm/18 in tall, so look good in a flower border. The roots can be left in the ground throughout the winter and when they start to sprout new leaves in the spring, they can be blanched in the ground and eaten cooked or raw in salads.

Salsify is closely related to scorzonera, which is a popular vegetable in Germany and can be used in the same way. It is more widely available in seed catalogues in the UK. Scorzonera differs from salsify in that it has yellow rather than purple flowers, broader leaves and a black-skinned root. Both vegetables are hardy and easy to grow and not prone to pests and diseases.

Varieties

o **Salsify Giant:** Long white roots with a distinctive flavour likened to oysters and asparagus.
o **Scorzonera Russian Giant:** Dark-skinned roots whose flavour improves after the first frost.

Cooking and preparing

Both salsify and scorzonera can be simply sliced and cooked as a vegetable and served with butter, or baked in a white or cheese sauce or fried in butter, garlic and parsley after boiling. These roots can also be added to soups and casseroles and deep fried like chips. All the plant can be used, blanch the sprouting leaves in spring by drawing up soil around them as they grow and cut them when they reach 15 cm/6 in high – eat them raw or lightly boiled or steamed. If the plants run to seed in dry weather, steam the young flower buds for a tasty side dish.

o **To prepare:** Peel, slice or cut into sticks and immediately plunge into cold water with some lemon juice added to prevent discoloration. Alternatively, boil with the skins on and rub off once cooked.

o   **To boil:** Bring a pan of lightly salted water to the boil and cook the prepared salsify or scorzonera for 15–20 minutes, depending on the size of the pieces, until tender.

Storing 🖼️

Salsify and scorzonera can stay in the ground all winter, but if the weather threatens to freeze them into the ground, dig up however many roots you may need, dry them and store layered in boxes in a cool, dark place.

Freezing ❄️

Choose young firm roots, scrub them and blanch them whole for 3 minutes. When cool, peel and cut into thick matchsticks, and pack into freezer bags and freeze for up to 12 months. Cook from frozen in boiling, lightly salted water for about 6 minutes.

SPINACH ❄️

Spinach is so easy to grow that it is a wonder that it is not more cheaply available in the shops. All the more reason to grow it yourself so that you can harvest the leaves young and small for salads (so expensive to buy) and large and mature for a nutritious cooked vegetable that's so much tastier and sweeter than those tins of spinach that Popeye used to explode into his mouth! Some varieties of spinach can be harvested through the winter, so there is no need to store it, although you can freeze it.

Varieties

There are several types of spinach – one that is harvested in the summer, another hardier type that crops through the autumn into winter and two more that aren't a true spinach – New Zealand spinach that looks very similar but with smaller, milder leaves, and perpetual spinach with is really a leaf beet, part of the beetroot family. There are many new varieties that are resistant to mildew and bolting.

o   **Perpetual Spinach (Leaf Beet):** A good type for dry soils that crops continuously throughout the summer and winter if you keep picking. Tastes and cooks just like true spinach.

S

o **Sigmaleaf:** A true spinach summer variety that can go on
cropping for a long time without bolting. It can also be sown
in the autumn for a winter crop.

o **Norvak:** A high-yielding summer variety that is slow to bolt.

o **Broad-leaved Prickly:** A standard winter variety with dark,
succulent green leaves. The seeds are prickly, hence the
name.

Cooking and preparing

Pick the leaves young and small and add raw to mixed salads or slice
the larger leaves and stir them into risottos and pastas for the last
few minutes of cooking, or serve steamed and mixed with a pinch
of grated nutmeg and cream. Combine lightly cooked spinach with
a mixture of crumbled feta cheese, ricotta and eggs and sandwich
between filo or puff pastry for Spanakopita, or Greek spinach pie.
Alternatively, serve cooked spinach with smoked haddock and a
poached egg for a quick nutritious supper.

o **To prepare:** Wash small, young leaves well, dry and add
whole to salads. Wash older, larger leaves well and slice them
wet into a large empty pan, including the tender stalks,
which are sweet when cooked. Using scissors instead of a
knife means that the spinach drops straight into the pan
without losing much water.

o **To cook:** Sprinkle the spinach with a little salt, stirring it
through the leaves and bring slowly to the boil, covered.
Cook for 5–8 minutes, shaking the pan occasionally, until
the spinach has wilted and shrunk to half its volume. Tip
into a sieve to drain, squeezing out as much moisture as
possible with the back of a spoon. Always use freshly picked
spinach straight away; don't leave it in the fridge for days as
it will become bitter and unpalatable.

S Freezing ❄

Young spinach leaves freeze well; large old, fibrous ones will have a strong taste and may be stringy.

- o **To prepare:** Remove any tough stalks and wash under cold running water. Drain well.
- o **To freeze:** Blanch the whole leaves for 1 minute, drain and squeeze out as much moisture as possible (see above). Chop or leave whole and put into freezer bags and freezer.
- o **To thaw and use:** Cook gently from frozen in a covered saucepan with a little salt but without any extra water for about 5 minutes, stirring occasionally, until piping hot.
- o **Storage time:** 12 months.

SQUASH & PUMPKINS ▣ ☼ ❄ ▤

American Indians were eating squash centuries before the Pilgrim Fathers arrived in America and made it one of their staple foods. There are two types of squash: summer squash, which are harvested young before their seeds harden, and include courgettes (page 122); and winter squash, including the pumpkin, which are hard-skinned and store well when they are mature. Their seeds can also be dried. Squash come in all shapes and sizes, and colours vary from pale cream and blue through to green and orange stripes to bright orange. They are easy to grow in a rich, well-watered soil and many varieties are trailing so look impressive draped around the vegetable patch.

Varieties

- o **Crown Prince:** A trailing winter squash with an attractive grey-blue shell and bright orange flesh with a sweet nutty flavour. The large fruit can grow up to 4 kg/9 lb in weight and can keep for many months.
- o **Avalon:** A wonderfully rich and sweet-flavoured trailing butternut that will keep for up to 6 months.
- o **Hasta La Pasta F1:** A great winter spaghetti squash that produces a good crop of oval, pale yellow fruits each weighing up to 1 kg/2¼ lb; inside, the spaghetti-like flesh makes a refreshing, healthy alternative to spaghetti.

S

- o **Pattypan 'Scallop Mixed':** A vigorous summer squash producing an unusual flat yellow or white fruit with a scalloped edge on a bushy plant. The small, young fruits can be eaten raw or cooked as you would courgettes (page 123). The fruits can reach 15–20 cm/6–8 in in diameter once mature.
- o **Mars F1:** A trailing pumpkin that produces good 2.75 kg/ 6 lb fruits that turn orange when ready to harvest. Stores well throughout the winter.

Cooking and preparing

Summer squashes can be sliced and cooked in butter, added to casseroles, made into soups, roasted in the oven and stuffed. Cook winter squashes with the skin on, so it can be removed easily when it is soft and pliable. Cut into wedges, lay skin side down on an oiled tray and bake until soft and slightly charred. The skins should come off easily. Chop the flesh into cubes and add to pasta dishes or risotto, or purée into a sauce, soup or mash. The smaller fruits can be stuffed. Roasting brings out the natural sweetness and intensifies the colour.

Storing 🖼️

Summer squash need to be eaten immediately but winter squash will store hanging in a net, or on shredded newspaper in a cool, airy, dry place for up to 6 months; some have been known to keep for a year. It is important to leave the fruit on the plant for as long as possible until they have matured properly. When there is a risk of colder, damper weather in October, harvest the fruits with some stem and move to a warm room indoors for a couple of days to help set the skin, then transfer them to their storage place.

Drying ☀️

Pumpkin seeds are very nutritious and can be dried on trays in a low oven. To remove the shells, toast them after drying with some salt. Some seeds have hulls that can be rubbed off easily.

Freezing ❄️

It is only worth freezing cooked pumpkin or squash. Bake them as above, then peel and either chop into cubes or mash. Pack into

S rigid containers and freeze for up to 12 months. To use, steam the frozen cubes over a pan of boiling water until heated through but thaw the mash at room temperature and use as required.

Preserving 🖾

Pumpkin makes an interesting jam but, as it has no pectin, add the juice and zest of 2 lemons. Also, because of its high water content, macerate in the sugar (equal weight of sugar to pumpkin) overnight before boiling. Add ground ginger, a stick of cinnamon or other spices according to your taste, if liked (see Marrow Jam, page 124).

Pumpkin and other squashes make a tasty chutney (see Marrow Chutney, page 125) and before cooking in vinegar, such as cider vinegar, the chopped flesh should be macerated overnight in salt, then drained and rinsed in cold water and drained again. Add onions and spices as you choose.

STRAWBERRIES ❄ ◗ 🖾

The best flavoured strawberries come from the garden and ideally they should be eaten freshly picked, but you should always save some to make strawberry jam – it is a true sign of summer when the aroma of strawberries cooking can be smelt wafting out of the kitchen. Strawberries are easy to grow, even in pots on the patio (although you do have to keep an eye on the birds and slugs), and nowadays there are so many varieties that the strawberry season can go on and on.

Varieties

Early summer fruiting varieties crop from mid-June to early July, mid-season varieties a week later. Late fruiting types begin to crop in July. Perpetual varieties produce a small amount of fruit in summer with the bulk from mid-August to mid-October.

o **Sonata:** High-yielding, sweet and juicy mid-season variety cropping from June to late July. Can cope with hot, dry periods and spells of heavy rain.

o **Malling Opal:** An early summer variety that produces heavy yields of large, sweet fruit right through to October.

o **Albion:** Disease-resistant heavy cropper that produces up to 450 g/1 lb delicious sweet strawberries per plant a year.

Fruits from June to October in several waves and does well in a patio planter.

o **Irresistible:** As the name implies, this strawberry is so sweet and juicy it is irresistible. A tender fruit, it is an ideal variety for the garden and containers, harvesting throughout the summer months.

Cooking and preparing

Strawberries should be eaten freshly picked, but if any make it to the kitchen you will never tire of them in desserts: pile them on to cream-filled pavlovas, or into almond pastry tarts and glaze with strawberry jam; sandwich them with cream between layers of shortcake or sponge cake; or purée them and turn them into mousse, ice cream, a soufflé or sorbet; or just eat them plain and simple with cream and sugar, or freshly ground black pepper and a splash of balsamic vinegar – the choice is yours. Only cook strawberries if they have been frozen.

o **To prepare:** Wash only if necessary, hull and use.

Freezing ❄

Strawberries lose their form and colour once frozen, so it is best to freeze them as a purée and then use it to make ice cream, mousses, etc. Purée the prepared strawberries in a blender or mash with a fork, add sugar to taste, if you like, and freeze in rigid containers for up to 12 months.

Juicing ◊

Strawberries are high in Vitamin C, so juice freshly picked fruit to enjoy all the health-giving benefits. They make a strong, thick juice that can be diluted with apple juice, citrus fruit juices or water. You can also make a strawberry cordial to pour over ice cream or to dilute with sparkling water for a refreshing drink.

S Preserving 🖾

Strawberries have very little pectin so strawberry jam needs to have the juice of 1 lemon added to it. Just leave the strawberries and sugar (in equal weights) to macerate for a few hours, stirring occasionally, and bring to the boil slowly with the lemon juice to dissolve the sugar. Boil for 5 minutes until setting point is reached.

SWEDE 🍲 ❄

The large, hard, round swede with orange-coloured flesh belongs to the same family as the turnip (page 203) and in Scotland is referred to as a neep, traditionally eaten mashed with haggis, so it can be confusing. Swedes are slower growers than turnips but they are hardy and can tolerate a cold northern winter in the ground, and store well in the cool and dry. With a deep purple skin fading to a cream colour at the base, swedes are a delicious but undervalued vegetable that will bolster up a hearty winter stew.

Varieties

o **Brora:** A tasty, sweet swede that is not prone to bitterness and can be harvested from September into New Year. Stores well.
o **Invitation:** A vigorous disease-resistant swede that is very winter hardy.
o **Marian:** A fine-flavoured, disease-resistant root.

Cooking and preparing

Individual swedes can grow up to 450 g/1 lb in weight, but are sweeter if they are harvested when they are smaller. Boil and mash them with plenty of butter and freshly ground black pepper – bashed neeps in Scotland – or mash them with floury potatoes and a finely chopped shallot or chives for the traditional Burns Night Orkney dish called clapshot, or champ in Northern Ireland, colcannon in the south. Swedes complement carrots well and can be served glazed together or deep fry them as you would potato chips. Add them to winter stews, soups and casseroles.

o **To prepare:** As the larger roots are hard to get a knife through, don't try to cut them in half as you may do yourself an injury. Slice off the sides first and cut them into chunks or dice, peeling as you go.

S

o **To cook:** Boil 2.5 cm/1 in chunks or slices in boiling, salted water for 15–20 minutes until tender, alternatively steam for 20–25 minutes.

Storing

Swedes keep well in the ground through winter but if you want smaller specimens, or the weather threatens to freeze the ground hard so that you are unable to harvest them, dig them up, wash, dry and store them layered in sand, or similar, in boxes in a cool, dry, dark place.

Freezing

o **To prepare:** Select small or medium-sized roots and peel and cut them into dice.
o **To freeze:** Blanch for 2 minutes, dry and freeze in freezer bags or rigid containers.
o **To thaw and serve:** Cook from frozen in boiling, lightly salted water for about 6–8 minutes until tender.
o **Storage time:** 12 months.

SWEETCORN

Growing your own sweetcorn is better than buying it as you can eat it fresh when the kernels are still packed with sugar and have not had the chance to turn into starch. So it is important that, if you do happen to have any surplus, to freeze it as quickly as possible. As the tall stems only produce one to three cobs on them, you need quite a few, planted in blocks to aid fertilisation, so plenty of space is important.

Varieties

Do not mix varieties as you will get a poor-tasting crop.

o **Conquest:** A reliable sweetcorn with early crops.
o **Extra Tender and Sweet:** A quick germinating variety said to be superior in taste and more vigorous than others. Produces 2–3 18 cm/7½ in cobs per plant.
o **Indian Summer:** A supersweet type that is high in sugars and produces red and purple kernels among the yellow ones.

S Cooking and preparing

Sweetcorn kept in its husk remains fresh for longer, protects the juicy kernels from bruising and prevents the cob drying out. Inside, the kernels should be tight, plump and soft buttermilk yellow. You can bake the whole cob, including the husks and silks, in an oven or on the barbecue, which will make the husks easy to remove. Cook husk-free cobs in deep boiling water for 15 minutes until the kernels loosen easily, or wrap individually in foil and bake or barbecue until tender. Resist adding salt during cooking as this toughens the kernels. Serve with generous amounts of melted butter. Great fun to chew off the cob, so serve whole or chop into chunks after cooking when the core is soft.

o **To prepare:** Remove the outer husks and silky fibres. If you want to use only the loose kernels, stand the cob upright on its base and run a sharp knife down the core, slicing the kernels off cleanly.

Freezing ❄

It is not worth freezing just the kernels, it is labour intensive as you need so many to make a bagful. However young, fresh cobs freeze beautifully.

o **To prepare:** Select cobs with plump juicy kernels and remove the husks and the silks.
o **To freeze:** Blanch for 5 minutes, dry well and seal in freezer bags and freeze.
o **To thaw and serve:** Cook from frozen in boiling water, with no salt, for 10–15 minutes until the kernels loosen easily. Drain and serve with melted butter, a little salt and freshly ground pepper. Alternatively, cook with a very little water in a covered dish in the microwave for about 3–4 minutes a cob, or according to your microwave instructions.
o **Storage time:** 12 months.

Preserving ▨

Sweetcorn kernels can be added to piccalilli (page 114) and made into a delicious relish, traditionally eaten with beefburgers. This one will keep for 3 months in a cool place.

— Sweetcorn Relish—

S

Makes about 550 g/1¼ lb

4 sweetcorn cobs
1 onion, finely chopped
1 fresh red chilli, finely chopped
1 red pepper, finely chopped
1 green pepper, finely chopped
Grated zest and juice of 1 large lime
275 g/10 oz clear, runny honey
400 ml/14 fl oz white wine vinegar
10 ml/2 tsp cornflour
Salt and freshly ground black pepper

1 Remove the husks and silks from the cobs and slice off the kernels. Tip them into a large pan of unsalted boiling water. Cook for 2–3 minutes until tender and drain well.

2 Bring the rest of the ingredients, except for the cornflour and salt and pepper, to the boil in a pan and cook for 15 minutes.

3 Stir the cornflour into 10 ml/2 tsp of water in a bowl to make a paste and add to the mixture with the cooked sweetcorn. Return to the boil and simmer for another 5 minutes until thickened. Season with salt and pepper to taste.

4 Spoon into warmed, sterilised jars and allow to cool. It can be eaten straight away.

S SWISS CHARD ❄

Swiss chard is not only a tasty and useful crop to grow, but it is also decorative, too, with plants that have bright pink, yellow, orange or white stems topped with thick, glossy dark green leaves and so look great in the flower border as well as the vegetable patch. The leaves taste and behave like spinach (page 188) in that they only need a little water to cook, and will continue to crop throughout the winter into the spring, so there is no need for storage. The stalks can be cooked separately as another vegetable.

Varieties

o **Bright Lights:** A fabulous mixture of different-coloured chards with stems of white, red, yellow and orange. Delicious to eat and lovely to look at.

o **Lucullus:** A white-stalked chard that produces a heavy yield. The white stalks are even sweeter and tastier than the coloured ones.

Cooking and preparing

Swiss chard is a luscious, easy-to-grow green with juicy leaves and vibrant stalks, however, the colours can bleed during cooking, tinting your dish. Cut away the leaves from the fleshy stalks and cook separately in different ways. Chard leaves have a slightly bitter earthiness and coarse texture. Cook and use as you would spinach, but remember it is more robust in flavour. Crunchy stems can be chopped and sautéed in butter and garlic or braised in wine as a sumptuous side dish or baked in a cheese sauce.

o **To prepare:** wash the leaves well, fold the leaves in half lengthways and cut out the stalk and chop into 2.5–5 cm/ 1–2 in pieces. Shred the leaves or cut with scissors.

o **To boil:** Bring a small quantity of salted water to the boil (about 1 cm/½ in deep) and cook the stalks for about 3 minutes, then add the leaves and continue boiling for another 3–5 minutes until reduced and just tender. Drain well.

Freezing ❄

As Swiss chard continues to grow outside throughout the winter there is not much call for freezing, and in any case the white stalks tend to go grey and look unappetising when frozen. However, you can freeze the stalks already baked in a cheese sauce.

o **To freeze:** Tie coloured, whole stalks in a bundle and blanch for 2 minutes, dry and freeze in freezer bags. Blanch the leaves whole for 1 minute, dry, shred and freeze in freezer bags. Freeze stalks baked in a cheese sauce in a covered foil container.

o **To thaw and use:** Cook leaves and stalks from frozen separately in boiling, lightly salted water for 6–8 minutes. Thaw the baked stalks, uncovered, at room temperature and heat through in the oven preheated to 180°C/350°F/gas 4 for 20 minutes until piping hot.

o **Storage time:** 6 months.

T TOMATOES ☀ ❄ ◐ ▤

Tomatoes are usually the first crop that an aspiring vegetable gardener will grow, although, strictly speaking, the tomato is a fruit. They can be grown in large pots on balconies and patios, hanging baskets, sheltered vegetable plots, and easiest of all in a greenhouse. Tomatoes can be very contrary: if they succeed then you will have a bountiful supply of flavoursome tomatoes, the like of which you cannot buy in Britain; but if they don't have enough sun, heat, food or water, or just too much of the latter, the fruit will turn brown or become tasteless.

Fresh from the vine, tomatoes have a heady aroma and an irresistible combination of sweetness, acidity and juiciness. They come in a wide variety of shapes and sizes, from tiny cherry tomatoes to the whopping beefsteaks and can be used fresh in salads and cooked in sauces, tarts and chutneys. Their flavour intensifies when dried and they make a very healthy, tasty juice.

Varieties

o **Floridity:** Tiny sweet plum-shaped tomatoes that can produce up to 35 fruits to a truss. Very high yielding under glass or outdoors. Freezes well.

o **Roncardo F1 Hybrid:** A beefsteak tomato producing fruits that weigh up to 120–140 g/4–5 oz each with very sweet flavoursome flesh that ripen early.

o **Hundreds and Thousands:** A vigorous, easy-to-grow cascading plant that lives up to its name producing a prolific crop of sweet, juicy cherry tomatoes. Perfect for hanging baskets, window boxes, or raised containers on the patio.

o **Cossack F1:** A medium-sized, tasty, high-yielding tomato suitable for outdoor growing in a sunny, sheltered spot. Good disease resistance.

o **Ferline:** A large heavy-cropping, deep red, flavoursome tomato with good resistance to tomato blight, a destructive fungal disease. Suitable for outdoor growing.

o **Pomodora Roma Nano:** Plum tomato that has more flesh and fewer seeds. Develops a fuller flavour if grown outdoors in a sheltered sunny spot. A good variety for drying.

Cooking and preparing

Partner lush ripe tomatoes with fresh basil leaves, mozzarella cheese, olive oil and plenty of seasoning. Give it some dash by combining different varieties, contrasting shape, colour, texture and flavour. Halved cherry tomatoes mixed with cubes of cooked beetroot, cubes of feta cheese, fresh coriander, lemon juice and olive oil make an unusual but colourful salad; slice beefsteak tomatoes and sprinkle with olive oil and thinly chopped spring onions for another tasty salad.

Intensify the flavours by grilling, roasting or baking tomatoes and then make into purée, soups (cold or hot) and sauces. Tomatoes are so versatile that not only can you stuff them or slice them on to tarts but they enrich every dish they are cooked in.

o **To prepare:** Tomatoes are usually best skinned before cooking, but it is not always important. Make a cross cut at the base of each tomato and plunge them into boiling water for a couple of minutes. The skins will start to loosen and peel off. Halve and scoop out the seeds if necessary.

Drying ☼

In a tropical climate tomatoes benefit from being dried in the sun, but that's not usually an option in the UK as they can rot quickly. Cut the tomatoes in half and either dry them in a solar dryer (page 24) or on a baking sheet very slowly in a very low oven. When dry, store them in a jar of olive oil with dried herbs (page 21). Eat them straight from the jar or use them to enrich soups stews and sauces.

Freezing ❄

Once frozen, tomatoes cannot be used raw in salads because their high water content makes them go mushy when thawed. However, they can be frozen raw to be used in cooked dishes or they can be frozen as purée, juice, roasted or cooked sauce.

o **Whole:** Wash and remove the calyxes of whole tomatoes, pack in a freezer bag and freeze. Thaw at room temperature for about 2 hours or longer in the fridge, then grill, fry or add to cooked dishes. They can be stewed from frozen.

o **For cooked sauce:** Make in your usual way, cool quickly and freeze in rigid containers. Use from frozen.

o **Storage time:** 12 months.

T

Juicing ▢

Peel off the skins (page 201), chop roughly and purée in a blender or food processor. Simmer for 5 minutes, push through a sieve and when the juice is completely cold, freeze in rigid containers. Thaw in the fridge overnight and serve cold. For purée, return the juice to the pan and continue to cook for about 30 minutes, stirring frequently to reduce the liquid and to thicken it to a purée. Cool quickly and freeze in ice cube trays or rigid containers. Each cube is equivalent to 15 ml/1 tbsp and can be added, frozen, to savoury dishes.

Preserving ▤

Make green tomato chutney from those tomatoes that just won't ripen and turn red, or red tomatoes into a chutney with apples, sultanas and onions. Process the result into a sauce and keep in sterilised bottles as described on page 41.

── Tomato Ketchup ─────────

Makes about 3 kg/7 lb

2.75 kg/6 lb ripe tomatoes
1 kg/2¼ lb onions
60 ml/4 tbsp salt
2.5 ml/½ tsp pepper
30 ml/2 tbsp mustard
1 kg/2¼ lb golden syrup
600 ml/1 pint vinegar

1 Skin the tomatoes and cut them in half. Peel and slice the onions.
2 Put them into separate bowls and sprinkle both with the salt. Leave to stand overnight.
3 Drain and put into a pan with all the other ingredients. Bring to the boil and simmer for 1¼ hours.
4 Put through a sieve. Reheat, pour into bottles or preserving jars, seal and sterilise (page 41).

── Tomato Jam ──   T

Tomatoes also make a delicious sweet jam

Makes about 3.5 kg/8 lb

2.75 kg/6 lb ripe tomatoes
Juice and grated zest of 6 lemons
2.75 kg/6 lb sugar
5 ml/1 tsp salt
10 ml/2 tsp ground ginger

1   Skin and halve the tomatoes, scoop out the seeds and chop the
    flesh. Put the flesh into a pan with the lemon juice and zest and
    cook gently until the mixture is reduced to a pulp.
2   Remove from the heat and gently stir in the sugar, salt and ginger.
3   Return to a low heat and stir until the sugar has dissolved. Boil hard
    to setting point.
4   Pour into warmed jars, seal and label.

TURNIPS 🎲 ❄️

Turnips are quick to grow, seeds sown in March or April can
produce a crop by May. Like salads, they can be sown in succession
throughout the growing season and harvested when they get to golf
ball size and eaten raw like radishes. Seeds sown in the autumn
produce a good crop of leafy tops in the spring that are delicious to
eat in salads, too. The part of the turnip that is eaten is not a root
but the swollen base of the stem. It is a member of the cabbage
family that includes the cauliflower, swede and kohlrabi and stores
well over winter.

T Varieties

- o **Aramis:** A purple-topped turnip ideal for growing close together to produce small roots with a nutty flavour. Will remain in the ground for weeks without getting tough.
- o **Armand:** A late-maturing purple and white turnip that can stay in the ground throughout winter.
- o **Tokyo Cross:** A flavoursome white turnip ready to harvest in as few as 35 days from sowing.

Cooking and preparing

Harvested young and small, turnips have a sweet flavour and sliced thinly are delicious raw in salads or steamed as a vegetable, or boil them whole in some orange juice for a tasty side dish. Later in the season their earthy, stronger flavour bolsters stews and soups; they absorb fat well so complement fatty meats, such as duck, pork and goose. Boiled turnips are also delicious puréed with some butter and nutmeg, or spices such as ground cumin and fenugreek. Roast them with parsnips and potatoes or bake them pommes Anna style. Their sprouting leaves and stalks can be used in salads or steamed as a vegetable accompaniment.

- o **To prepare:** Scrub turnips and trim well, peeling off the skin if tough. Wash and slice the leafy tops.
- o **To cook:** Steam young turnips for 8–10 minutes, for larger turnips, cut into quarters and boil in lightly salted water for 20–25 minutes. Steam the leaves for about 5 minutes.

Storing 🔲

Select firm, unblemished turnips, scrub off the soil and dry well. Layer in boxes of sand, or similar, and store in a dark, cool, dry place. Cut off and use the leaves when they sprout.

Freezing ❄

- o **To prepare:** Select small or medium-sized turnips. Wash and leave the small ones whole, peel and chop the larger ones into dice.
- o **To freeze:** Blanch for 2 minutes, dry well and freeze in freezer bags.
- o **To thaw and use:** Cook from frozen in boiling, lightly salted water for about 6–8 minutes until tender.
- o **Storage time:** 12 months.

Recipes in *italic*

apples 68–74
  apple & ginger pear juice 162
  apple & plum butter 71
  apple aniseed 131
  blackberry & apple jam 94–5
  sparkling apple wine 72–4
  storage 16, 70
  traditional apple chutney 72
  see also cider
apricots 74–6
artichokes, globe 76–8
artichokes, Jerusalem 79–81
  Jerusalem artichoke soup 80–81
asparagus 81–3
  asparagus tart 82
aubergines 84–5
  baba ghanoush 85

bagging up 16
beans, French 86–8
beans, runner 88–90
  classic bean chutney 89–90
beetroot 90–2
  beetroot muffins 91
  beetroot wine 92
  preserving 92
blackberries 93–6
  blackberry & apple jam 94–5
  blackberry liqueur 96
  blackberry wine 95
  spicy blackberry chutney 95
  sugar-free blackberry jam 50
blackcurrants 127–9
  blackcurrant jam 129
  blackcurrant wine 129
blanching 29, 30–31

blueberries 96–9
  blueberry jam 98
  blueberry pie 98–9
bottling 39–44
  equipment 40
  methods 41–4
  preparation 40–41
boxes, storage 15, 16
brining 53
broad beans 99–101
  broad bean wine 100–101
broccoli 101–2
brussels sprouts 103–4

cabbages 105–8
  sauerkraut 107–8
cakes
  carrot & apple cake 110–11
  cherry lemon cake 119–20
carrots 108–11
  carrot & apple cake 110–11
  carrot wine 111
cauliflower 112–14
  piccalilli 114
celeriac 115–16
celery 116–17
cherries 118–20
  cherry lemon cake 119–20
chicory 120–2
chilli peppers 166–8
chutneys 51, 54
  classic bean chutney 89–90
  fig chutney 133
  green plum chutney 170–71
  marrow chutney 125
  muscovado rhubarb chutney 186
  Peggy's pear chutney 163
  spicy blackberry chutney 95

spicy gooseberry chutney 138
traditional apple chutney 72
cider 55–6
equipment 59–61
making 59, 66, 72
clamps, storage 15, 18
compost 6
cook and freeze method 36–8
cordials 55–6, 58
courgettes 122–5
marrow chutney 125
marrow jam 124–5
marrow wine 125
crop rotation 10
cucumber 126–7
currants (black, red & white) 127–9
blackcurrant jam 129
blackcurrant wine 129
redcurrant jelly 129

damsons
pickled damsons 172
dry storage 13
drying 19–26
equipment 20

fennel 130–31
apple aniseed 131
figs 132–3
fig chutney 133
fig jam 133
flowers, drying 22
freezers 28, 29
freezing 27–38
cook and freeze method 36–8
dry packing 31
equipment 28
methods 30–5
in syrup 31
and thawing 37–8
freshness 9

frost protection 14, 17
fruit
drying 23–4
preserving in alcohol 44
fruit butters 45, 49
apple & plum butter 71
fruit curds 45, 49

garlic 134–6
gazpacho 135
pickled garlic 135
storage 17, 135
gooseberries 137–9
gooseberry jam 138
gooseberry wine 139
spicy gooseberry chutney 138
grapes 139–42
see also wine
ground, storing in 15, 17
growing your own food 5–6

hanging up 15, 17
harvest 9, 12
herbs 143–5
bajan seasoning 144
drying 19, 21–2, 145
storage 22

jam 45, 47–8
blackcurrant jam 129
blueberry jam 98
fig jam 133
fruity raspberry jam 182
gooseberry jam 138
marrow jam 124–5
rhubarb & rose petal jam 185
sugar-free blackberry jam 50
tomato jam 203
jars (glass) 40, 47, 52
jellies 45–7
making 48

quince jelly 178
redcurrant jelly 129
juicers 55–6
juices 55–9

kale 145–6
kohlrabi 147

leeks 148–50
leek gratin 150

manuring 10
marmalade
red onion marmalade 154
marrows
marrow chutney 125
marrow jam 124–5
marrow wine 125

nectarines *see* peaches

onions 151–4
red onion marmalade 154
storage 17, 152

parsnips 155–7
parsnip wine 156–7
peaches 157–9
pears 159–64
apple & ginger pear juice 162
pears in Burgundy 161–2
Peggy's pear chutney 163
perry 163–4
storage 16, 160
peas 164–5
pea & mint soup 165
pectin 46
peppers 166–8
pests 14
pickles 51–3
pickled damsons 172

pickled garlic 135
pickling equipment 52
pies & tarts
asparagus tart 82
blueberry pie 98–9
raspberry tart 180
planning 8–11
plums 168–72
apple & plum butter 71
green plum chutney 170–71
pickled damsons 172
plum sauce 171
plum wine 172
potatoes 173–5
cooking & preparing 174
preserves
preserving equipment 47
see also individual types
pumpkins 190–2
purées 31

quinces 176–8
quince jelly 178

raspberries 179–82
fruity raspberry jam 182
raspberry tart 180
raspberry vinegar 182
redcurrants 127–9
redcurrant jelly 129
relishes
sweetcorn relish 197
rhubarb 183–6
muscovado rhubarb chutney 186
rhubarb & ginger crumble 184–5
rhubarb & rose petal jam 185
stewed rhubarb 184
root vegetables 16, 18
see also individual types

salsify 187–8

sauces 54
  plum sauce 171
seeds, drying 22
shallots 17, 151–4
solar dryers 24
soup
  gazpacho 135
  Jerusalem artichoke soup 80–81
  pea & mint soup 165
spinach 188–90
squash 190–2
storage 12, 13–18
  conditions 14
  dry storage 13
  equipment 15
  space 9
strawberries 192–4
swedes 194–5
sweetcorn 195–7
  sweetcorn relish 197
Swiss chard 198–9

tainting 14
thawing 37–8
tomatoes 200–3
  tomato jam 203

  tomato ketchup 202
trays 15, 16
turnips 203–4

vegetables
  drying 25–6
  root vegetables 16, 18
  see also individual types
ventilation 14
vinegars 53
  raspberry vinegar 182

wine
  beetroot wine 92
  blackberry wine 95
  blackcurrant wine 129
  broad bean wine 100–101
  carrot wine 111
  gooseberry wine 139
  marrow wine 125
  parsnip wine 156–7
  plum wine 172
  sparkling apple wine 72–4
wine making 59, 62–5, 142
  equipment 59–61

Useful websites